The Mantra of Efficiency

The Mantra of Efficiency

From Waterwheel to Social Control

JENNIFER KARNS ALEXANDER

The Johns Hopkins University Press
Baltimore

The Johns Hopkins University Press
2715 North Charles Street
Baltimore, Maryland 21218-4363
www.press.jhu.edu

Library of Congress Cataloging-in-Publication Data

Alexander, Jennifer Karns, 1963–
The mantra of efficiency : from waterwheel to social control /
Jennifer Karns Alexander.
p.   cm.
Includes bibliographical references and index.
ISBN-13: 978-0-8018-8693-5 (hardcover : alk. paper)
ISBN-10: 0-8018-8693-7 (hardcover : alk. paper)
1. Technological innovations—History. 2. Industrial efficiency.
3. Science and industry. I. Title.
HC79.T4A443   2007
338'.06409—dc22   2007018866

A catalog record for this book is available from the British Library.

A previous version of chapter 5 appeared as "Efficiency and Pathology:
Mechanical Discipline and Efficient Worker Seating in Germany,
1929–1932," *Technology and Culture* 47 (2006): 286–310.

Special discounts are available for bulk purchases of this book.
For more information, please contact Special Sales at 410-516-6936 or
specialsales@press.jhu.edu.

To Jim

CONTENTS

Two of my own personal experiences of efficiency have greatly influenced this book. Both involved motion control, which has been central to efficiency. The first experience is my use of motion control orthotics to correct my gait when I had trouble walking in the early 1990s (having them made involved me being videotaped on a treadmill, my legs and feet marked up with orange ink). The devices revolutionized my life. They made it possible for me to hike to places my family had long gone without me and to take up jazz dance, but I experienced them as controlling. They were rigid and forced my feet to move in a much smaller arc of motion than I was used to, along a very narrow path. I resented having to wear them, and I did not like inefficiency being used to diagnose the pathological deformity of my feet.

The other experience was observing my young son's struggle with apraxia as he learned to control the muscles he needed for speech. His "speech efficiency," as his therapist called it, improved markedly when he learned to use physical cues to draw his attention to the muscles he needed to move. One of his first deliberately formed sounds was an explosive "buh," for the letter *B:* he had learned to make a fist and rest his chin in the hollow between thumb and knuckle, bringing his attention to his lower lip; then he jerked his fist away while expelling as much air as he could. It was a lovely sound. I had none of the qualms about the efficiency of his speech that I'd had about the efficiency of my gait; he had been miserable too long. This contrast, between the benefits and the costs of efficiency, between deep hurt and great good, lies at the heart of this book.

What I had come up against was a technological orthodoxy: the belief that all things should act efficiently. Like all orthodoxies it offers comfort and guidance, but, as orthodoxies do, it also has the power to wound those who cannot follow its dogmas or who resist its rituals of conformity. It is technological because it has primarily to do with making things work, and it is particularly apparent in the contemporary emphasis on quantifiable productivity and associated fears of waste, especially the waste of time. Historical study offers a tool for uncovering and critically examining the technological orthodoxies that increasingly dominate life in industrial and postindustrial societies.

Efficiency as a feature of modern industrial culture is the topic of this book. It begins with an introductory discussion of the historical varieties of efficiency, which suggests several methods of organizing the apparent and important contradiction between its use as a tool of conservation and a tool of growth. The chapters that follow are arranged chronologically, beginning with a case study from the early Industrial Revolution and ending with a discussion of whether efficiency, as an industrial value, will have a place in a postindustrial order. The individual studies may be grouped according to three historical themes: the early industrial history of output and input comparisons, which formed efficiency's mechanical and intellectual core; efficiency's formalization as a technical concept, alongside its early use in other fields, here biology and economics; and the widening use of the concept after the turn of the twentieth century, often in a time of turbulence.

Evaluating the effectiveness of machines in motion is the subject of the first two chapters. Celebrated British engineer John Smeaton's experiments on the power of waterwheels in the mid-eighteenth century and trials done by the American Franklin Institute nearly a century later are the subject of chapter 1. Both efforts compared work done by the machine with the power available in the water that drove it, and both agreed that controlling how the water moved through the machine was crucial to improving its effectiveness. In contrast, how they treated details of the moving machine, and underlying philosophical issues, differed significantly. The chapter asks how Smeaton and experimenters at the Franklin Institute conceived of the relationship between the machine's moving and static parts and how this conception influenced their widely publicized recommendations on how to design effective waterwheels. The relationship between an early-eighteenth-century idea of the perfected machine and the control of its motion is the subject of chapter 2. It analyzes the work of Gérard-Joseph Christian, director of the Paris Conservatoire des Arts et Métiers during the Bourbon Restoration. Christian developed a far-reaching philosophy of industrial society that turned on distinctions between how machines and people moved. The chapter asks how Christian's understanding of different types of movement influenced his concept of the most perfected machines, which he described as the most effective. Together, these chapters explore the historical relationship between efficiency, change, and motion in the concrete historical practice of machine design and construction.

Attention then turns to the formalization of efficiency as a concept in engineering and the beginning of its broader, more popular career. Its technical definition depended on the laws of thermodynamics of the mid-nineteenth century and was not complete until nearly fifty years later, but efficiency appeared in common use and in other technical forms long before it was fully formalized. Chapter 3 raises the issue of efficiency as an intellectual construction in other fields, particularly in Charles Darwin's theory of natural selection and in neoclassical economist Alfred Marshall's conceptions of the differing capacities for labor of various ethnicities and races. It asks how Darwin's and Marshall's uses of efficiency reflected the important and developing distinction between static conceptions of phenomena and dynamic ones and whether efficiency can be described simultaneously as a tool of conservation and a tool of growth.

The remaining chapters examine broadening uses of efficiency in the twentieth century and ask what similarities existed between them and between efficiency in its technical and common forms. Chapter 4 compares engineering conceptions of efficiency and its popular use in the Progressive-era United States by analyzing how it appeared in two different and widely circulating journals: a technical journal, *Engineering Magazine,* and the respected journal of reform, the *Independent.* Efforts to reduce the burdens of poverty, to improve conditions of industrial labor, and to remake political life characterized the Progressive era, and the pursuit of efficiency was central to many such efforts. The chapter asks what the relationship was between efficiency in engines and efficiency in social and personal use.

Efficiency, social crisis, and reform are the themes of the next two chapters. Chapter 5 returns to the relationship between efficiency and motion control as expressed in an exhibit on labor and posture that traveled through Germany in the crisis years before the Nazi seizure of power. The chapter asks what efficiency offered in a time of crisis and how a minor reform, efficient posture at work, was related to larger efforts to transform a society in crisis. Moral considerations of efficiency are considered in chapter 6, which analyzes the bitter controversy over *Time on the Cross,* the 1974 study by Fogel and Engerman that concluded American slavery had been an efficient form of business. Critics saw the argument as an apologia for slavery; the authors responded that African Americans should take pride in the efficiency of their ancestors. The chapter examines the continuing moral force behind conceptions of efficiency and looks particularly at Fogel's change of heart in the late 1980s, when he ceased to defend the efficiency of slavery as proof of slaves' good work ethic. It asks how efficiency and moral value came to be conflated in the debate and how one might treat differences in the experience of efficiency, between those who benefited from it and those who were exploited.

The final chapter considers the role of efficiency in contemporary times and asks which of its historical features will continue to resonate in a postindustrial and globalized world. Key is the concept of flow—of information or of goods, for example—and the role of efficiency in preventing disruptions. The conclusion considers the fundamental relationship between efficiency and effective human agency as expressed in the practice of prediction and planning. While the conclusion acknowledges that efficiency has taken on a wide variety of forms, it also suggests that beneath its multiformity is a strong and general adaptability, allowing it to be turned to many different uses and applied in widely divergent circumstances. It suggests that beneath the zeal for efficiency lies the desire to control a changing world, by bringing it into conformity with a vision of how the world does or should work.

ACKNOWLEDGMENTS

I owe a great debt to many persons and institutions that made it possible for me to write this book. Bruce Hevly at the University of Washington guided the project in its earliest stages, and he has read and reread manuscript drafts. He has been an unfailing source of intellectual stimulation and support. Thomas Hankins, Uta Poiger, and Elizabeth Domansky also greatly influenced the early development of the project. Colleagues in the Program for the History of Science, Technology, and Medicine at the University of Minnesota provided critical support. Arthur Norberg and Robert Seidel read the entire manuscript and provided extensive and insightful criticisms. Michel Janssen helped police my "mental hygiene," as he put it, by providing a useful and philosophically minded review; and I benefited greatly from long conversations with Jennifer Gunn. Richard Bellon, now at Michigan State University, provided an important critique of the section on Darwin. Colleagues in the Mechanical Engineering Department at the University of Minnesota—including Art Erdman, Marie Guion, David Kittelson, and Kim Stelson—advised me on important technical matters.

The anonymous reviewer for the Johns Hopkins University Press gave me necessary direction in bringing an unruly project into manageable shape. Bob Brugger, at JHU Press, has supported the project from its early stages. His laconic suggestions have been well timed and to the point, and at crucial junctures he provided very specific and welcome guidance. The editing of Mary Keirstead and Elizabeth Gratch improved the style and argument considerably. I also benefited from the comments and assistance of a number of people, including Erika Dirkse, Suzanne Fischer, Karen Freeze, Gary Fouty, Nina Lerman, Hiromi Mizuno, Don Opitz, David Sepkoski, J. B. Shank, Donna Shear, Eric Weitz, and Carol Zinda.

The University of Washington Department of History supported early research through a Rondeau Evans Dissertation Fellowship. A fellowship with the Centre de Recherche en Histoire des Sciences et des Techniques, Centre National de la Recherche Scientifique, under the direction of John Krige in Paris, enabled me to do much necessary work in French sources. The Center for German and European Studies (CGES) at the University of Minnesota supported much of the writing through a research collaborative with Tom Broman, of the University of Wisconsin, Madison. Tom was a delight to work with, energetic and irreverent. CGES also sponsored a round-table discussion of the manuscript in its late stages, which was decisive in bringing the book into its final form.

I owe enormous debts to the staffs of a variety of archives and research libraries. M. Premoselli at the archive of the Institut National de la Propriété Industrielle in Compiègne opened its vault for me, and the archives at the Conservatoire National des Arts et Métiers gave me access to beautifully preserved engravings. The staff at the Bibliothèque Nationale and the Archives Nationales were unfailingly helpful in retrieving many carefully filed documents and in finding a number of them that had been mislabeled. The Bundesarchiv in Koblenz and in Berlin kept me notified as records moved between them. Alex Saunders assisted me in the Marshall Papers at the Marshall Library of Economics, Cambridge University. Carol Arrowsmith assisted me with materials on John Smeaton in the archives of the Institution of Civil Engineers, London; and Mary Sampson, Archivist of the Royal Society, London, assisted me with materials there. The manuscripts division of the British Museum helped locate some of Smeaton's correspondence. The Department of Reproductions at the Staatsbibliothek, Berlin, assisted with photographs, as did the Harlan Hatcher Graduate Library of the University of Michigan and the Wangensteen Historical Library of Biology and Medicine at the University of Minnesota. Other photographs were taken and prepared by Nancy Johnson of the Department of Mechanical Engineering at the University of Minnesota.

I also owe a personal debt to several people without whom this project would never have been completed. Margaret Paton-Walsh set a high standard of scholarly and personal integrity, and it is a loss to history that she has left the profession. The members of Newton for Ladies—Jacqueline Ettinger, Lisa Hendrickson Nakamura, and Constance Malpas—became friends as well as critics. Moran Tompkins influenced me greatly, with good common sense and unfailing support throughout a long project.

Three acknowledgments remain. My mother, Phyllis Spear Karns, was a constant source of support and shrewd advice. Jack, my young son, arrived during the writing of this manuscript, and he has been a continual diversion and joy.

Jim, my husband, has long been my model of integrity and discipline, and as I finished this project, I truly learned how deep are his resources. He did not measure; he did not calculate; he did not ration. He gave, and gave again. I dedicate this book to him.

The Mantra of Efficiency

Introduction
The Varieties of Efficiency

What saves us is efficiency—the devotion to efficiency.
—*Marlow, in Joseph Conrad,* Heart of Darkness

Efficiency is celebrated. Joseph Conrad's Marlow paid homage to the foundations of the industrial culture he represented in the novella *Heart of Darkness* (1902), a culture forgotten and betrayed by his quarry, Kurtz, who had surrendered to savagery and wilderness in horrible pursuit of the "forgotten and brutal instincts" that Marlow believed efficiency had tamed. As Conrad told it, efficiency separated savage from civilized. Six decades later the economic historian David Landes found a similar role for efficiency in what would become a classic study of the Industrial Revolution. Landes cited the capacity to take deliberate and effective action as the mark of an advanced society, which he associated with science and progress and opposed to magic. Efficiency has indeed become a central value in the world's advanced industrial cultures, an apparently self-evident value associated with individual discipline, superior management, and increased profits. Conrad and Landes suggest a deeper meaning of efficiency as a moral good in and of itself, born of the same roots that brought longevity, wealth, and security to the earliest industrial nations of the earth.[1]

Efficiency is also a deeply troubling idea. Critics of industrial society have turned to it in concern rather than celebration, seeing in it not a harbinger of progress but a technique of control and exploitation. The French philosopher and lay theologian Jacques Ellul found the quest for efficiency antithetical to human freedom, believing it ultimately required all things human to be analyzed and integrated into orderly and manageable systems. American cultural critic Lewis Mumford rejected capitalist uses of efficiency, which he believed were tied to profit seeking and individual enrichment at the expense of true social and cultural advance.[2]

Enthusiasm for efficiency remains high; only in intellectual circles have critiques such as those by Ellul and Mumford been seriously engaged. People see in efficiency a solution to scarcity, a way to make less go further, and they summon it whenever a crisis of resources strikes. Advocates of privatizing U.S. prisons and schools offer efficiency as a justification, ever more attractive as money dwindles and more clients enter each system. Rising fuel prices and renewed worries about dependence on foreign petroleum have made efficiency as popular a term as it was during the oil crisis of the 1970s. Efficiency remains an iconic mantra in the high-tech industries, which continue to promise new and improved technologies to stretch the corporate dollar further and further. Although there may be situations in which people would question looking for efficiency (think of love and faith), in general people speak of efficiency as a good thing and associate it with a job well and economically done. All other things equal, better efficient than not.

This study asks how this valuation came to be, how an obscure philosophical concept—and that is how efficiency started out—became an important industrial tool and, later, a popular social and personal ideal. Efficiency was not always simultaneously celebrated and condemned. It was not always described as one of the foundations of civilization or a threat to human liberty, and it did not always speak of business, management, and profits. Until the nineteenth century efficiency remained a philosophical concept, associated with the causes of change and the ways of God, and only during the Industrial Revolution was it linked with human powers and abilities. Efficiency then became one of the foundational concepts of the modern industrial age. It was an industrial invention, created by engineers and physicists to measure the performance of machines, and, in particular, to relate a machine's output to the inputs it had used. Efficiency did not remain technical for long: an efficiency craze swept the United States and Europe in the first decades of the twentieth century and made *efficiency* a household word. Efficiency became promiscuous, describing activities of all sorts, including marriage, fuel consumption, use of leisure time, and political and moral behavior.

Its technological heritage gives efficiency a sense of objective plausibility, underlying the enormous variety of ways it has been used. Two distinct senses of the word stand out, with discrete but allied meanings, apparent even in the root physical activities that defined efficiency in engineering practice in the eighteenth and early nineteenth centuries. One was an efficiency of balance, a static efficiency, the highest measure of which accounted for the conservation of measured elements. The other was a creative and dynamic efficiency, which allowed growth through careful management and brought as its reward not merely conservation but growth. These two meanings were woven together in what appeared to be a paradoxical rhetoric stressing both the conservative character of efficiency and its dynamic and creative potential. Through the nineteenth and twentieth centuries the notion of efficiency gained credibility by the breadth of its applications: to biology, economic thought, personal development, worker management, and social history. Yet its bifurcated foundation remained, creating problems for any coherent application of the idea. Despite this ambiguity, modern culture elevated the concept, papering over its internal contradictions and the wide variety of ways it was applied to different problems and different people.

The growing power of efficiency, and its broad persuasive value despite its lack of a unified and coherent expression, is the subject of this book.

The Problem

Something happened to the idea of efficiency between the seventeenth and twentieth centuries. Efficiency had been a philosophical concept describing agents and causes of change, yet it dropped out of sight in the eighteenth century, only to resurface in the nineteenth in a different form, as a technical measurement of the performance of machines.[3] It moved into economics and then, early in the twentieth century, into more common use, as an efficiency craze swept through Europe and the United States. Its protagonists were men such as Frederick Winslow Taylor, founder of scientific management; Henry Fayol, French specialist in labor organization; Henry Ford, whom many Germans saw as the symbol of productivity and plenty; and Lord Roseberry, advocate of British political reform. Conferences on efficiency, labor organization, and productivity began to meet across Europe, peopled by representatives from nations struggling to industrialize: Czechoslovakia, Greece, Italy, and Poland.[4]

How to characterize efficiency, beyond recognizing its ubiquity, is a difficult problem. A central underlying issue is how to assess the modernity of the world of which efficiency was a part. That world is being left behind, succeeded by an age widely perceived to be new and called by a variety of names: postmodern, postindustrial, late capitalist. Two difficulties arise. First, there is the issue of efficiency's many uses, which suggest that it is an instrumental value, without inherent meaning of its own. Then there is the question of its role in industrial modernity, the historical period to which it belongs; but how to define modernity itself is a highly controversial problem, and the role of technology within it remains unclear. The temptation with multiple uses is to reduce efficiency to a matter of context; with the question of modernity, to subsume it in broad notions of Enlightenment rationality. Neither practice is in itself misguided, but the ground between them remains vast, between descriptions of efficiency in specific historical cases and its role in the broad sweep of modernity, however defined.

Efficiency is a slippery concept. As the previous examples illustrate, it has taken on not only a variety of technical configurations but also a bewildering array of more common meanings. At times it emphasizes stability and conservation; at other times it is embedded in a rhetoric of transformation. It appears to be merely a technique of quantification, yet it also appears as the goal toward which quantification is employed. It can be both the model of a well-controlled process and a tool to help achieve that control.

The temptation, when confronted with such a broad array of uses, is to treat efficiency in one of three ways: as the product of a specific localized context, as the metaphorical application of an engineering term, or as one among many instances of quantification. All three capture crucial aspects of the concept's character. The decision to control systems or processes through efficiency did reflect contexts of choice, for example, in national efficiency reforms in Britain at the turn of the twentieth century, when the empire found itself surprisingly challenged in the Boer War.[5] Efficiency is useful as a metaphor, especially when applied to processes outside human control; Charles Darwin developed an extended metaphor of the efficiency of natural selection in *The Origin of Species*. Technically, efficiency was indeed an exercise in quantification, measuring how much power or fuel a machine used and comparing that amount to how much work it did.

But these approaches fail to account fully for efficiency's historical complexity. Subsuming efficiency in context suggests that it is a shell, ready and waiting to take on the values and objectives of whoever uses it but with little content or character of its own. This interpretation has allowed scholars to manage the concept's widely disparate uses by centering discussion on the goals it served. Equally important, emphasizing context has allowed scholars to address complicated issues of moral value that often surround efficiency, chief among them its use in the service of inhumane or immoral ends. An example cited by scholars and students alike is the use of efficiency by Germany's National Socialists in creating a system of mass murder in the Holocaust.[6] The example is the elephant in the room in any discussion of efficiency as a value, and it is used to demonstrate the concept's role as a mere tool or instrument that takes its value only from the context in which it is used. The Nazi genocide was not a well-planned system, despite pockets of systematic and directed murder; in general it was a bloody, ugly mess.[7] It was quite effective nonetheless at destroying throughout much of Europe the Jewish peoples and cultures so hated by the Nazis, along with significant numbers of other peoples cruelly labeled degenerate, including homosexuals, Sinti, and Roma. Such brutal effectiveness is not necessarily the same thing as efficiency, but the example does illustrate the critical importance of context in assessing efficiency's historical meaning. It cannot be said to be an intrinsic value if it is primarily an instrument for implementing the other values that define the context of its use. Collapsing efficiency into its context thus reinforces the sense that it is a neutral technique.

Complicated situations in which efficiency was invoked as a moral value occurred repeatedly in the twentieth century. Two vignettes illustrate the complexity of the problem. Both linked efficiency with human freedom, among the highest and most contested of values in the liberal and industrialized West; both portrayed efficiency as freedom's opposite. Each scene grew out of deep social and political conflict, occasioned in the first case by the Great Depression and in the second by racial tensions in the civil rights era United States.

The first scene took place in Madison Square Gardens, New York, on October 31, 1932. President Herbert Hoover strode to the platform as the building shook to the cheers of more than twenty-two thousand supporters; thirty thousand more jammed the streets outside. At stake was freedom itself, as Hoover put it in this, one of the last speeches of his reelection campaign. The essential choice between himself and Franklin Delano Roosevelt was between freedom and bureaucracy, between decentralized and limited government relying on cooperation and voluntarism and a "new deal" that was also a new philosophy, extending government into business, banking, power production, and job guarantees. FDR's proposed New Deal promised "the maximum of efficiency" but only "at the cost of freedom."[8]

The second scene was a debate broadcast on the *Today Show* on April 30, 1974. Robert William Fogel, economist and future Nobel Prize laureate, appeared "visibly chagrined" by charges that he was defending the efficiency of American antebellum slavery in his new book, *Time on the Cross,* written with Stanley Engerman. Kenneth Clark, eminent sociologist, charged that Fogel's efficiency argument portrayed slavery as merely a "benign form of oppression." In dispute was whether Fogel and Engerman's precise mathematical measures of slavery's efficiency had overlooked its cruelty and, in the words of historian Kenneth Stampp, created "a tidy, rational world that never was."[9]

The complexity of these scenes arises not only from the juxtaposition of freedom and efficiency but because both invoked efficiency in ways meant to transcend the context in which they spoke. Hoover spoke to the future of a nation that had struggled to preserve its freedom for a century and a half, Fogel and Clark both to the past and the future, to the suffering of slavery's victims and the legacy carried by their descendants. Each was a single use of efficiency but colored by its multitude of other meanings. Hoover knew efficiency as a technical matter, as a specific and highly useful quantitative tool; he was a mining engineer by training and had been an engineer and manager by profession before becoming a politician. As president, Hoover also knew the popular appeal of efficiency; it had become a ubiquitous slogan, and not only in the United States, hailed as the solution to poverty, urban deterioration, and a host of other social and economic ills.[10] Fogel, too, knew the term in both its technical and popular senses. He and his coauthor had used complex mathematical calculations to compare the returns of slavery with its costs. But technical nuances were not the point in Fogel's confrontation with Clark; the point was whether efficiency itself was good, or, in other words, whether efficiency had a meaning that transcended its context.[11]

Treating efficiency as an engineering metaphor extends the idea of efficiency as context bound by differentiating between a sense in which its use is authentic and one in which it is derivative, in fields, for example, such as aesthetics or politics. Speaking of metaphor suggests that efficiency has no substantial content of its own when used outside of the physical, material domain of engines and energy.[12] Treating efficiency as a metaphor sidesteps the thorny question of how a technical concept might effectively be applied in nontechnical circumstances and undermines the possibility of recognizing serious attempts to extend a powerful and generalizable concept beyond the boundaries of physics or engineering.

Efficiency as an instance of quantification also raises important issues, especially concerning authority, trust, and expertise. The process of quantification is itself nuanced and relies both on additional techniques of standardization and measurement and ongoing social and political negotiations over who has authority to measure and by what standard. Numbers have come to carry a power of their own, backed by a machinery of expertise that produces authorities trained in interpreting them for audiences to whom they would otherwise remain unintelligible. Efficiency is one of many quantifying tools that reduce complex issues to mere numbers. It is also much more—for example, in the work on health, culture, and efficiency of British statistician and eugenicist Karl Pearson, for whom it described effective social reform.[13]

Context, metaphor, and quantification are thus important considerations in confronting the various meanings of efficiency, but they are unable adequately to describe its appeal across different contexts, its powerful application outside the engineering fields, or the transcendent rhetoric that portrays it as the wellspring of civilization or a threat to human freedom. These techniques have provided valuable insights into particular aspects of the concept, but they have also made it difficult to give serious attention to its broader meanings in modern industrial culture.

Efficiency is indeed a modern concept, having grown up alongside the industrial developments and newly systematic and rational processes that characterized an age well conscious of its emerging productive and administrative powers. It combined measurements of both the performance and resources used by machines, systems, or processes and, in its mature technical or mechanical forms, relied on thermodynamic laws describing the interconvertibility of energy, heat, and motion. It thus embodied the modern ideal of unified and testable knowledge, based on a fundamental, unimpeachable foundation. But beyond generalities the precise relationship of technology to modernity remains unclear. It has been central to analyses of the differences between recent centuries and the medieval past but most often appears as a monolithic, undifferentiated force rather than a varied and complex, even contradictory, working out of human skills and desires. The fact is, the great theorists of modernity have not found specific technologies important enough to address in detail.[14] The problem is not just that theoretical treatments of modernity have failed to engage technology as the historically complex subject it is. The problem is that this very complexity raises important issues at the foundation of conceptions of modernity, having to do with human reason and human emancipation.[15]

Calling something "modern" was more than a claim that it was new or up-to-date; historically, it was also a claim that the new or up-to-date was good. To be modern was to be open to continual change, in the belief that change was progressive and improving, even if it required the sacrifice of beloved traditions and values. The full sense of modernity as continual improvement came with the linking together of emancipation and rationality in the Enlightenment and the belief that humankind could use its own reason not only to challenge claims to authority but also to understand and ameliorate the human condition.

As an expression of Enlightenment rationality, modernity described the search for knowledge of epistemological certainty, useful not only for understanding the human and material worlds but also for directing their improvement. Central was the building of grand narratives of progress, built upon a positive belief in the human ability to uncover the single and unified foundational truths that underlay both the human and natural worlds. Technological developments seemed to support a modern narrative of progress: through the rational, disciplined application of natural laws, technology appeared steadily to have improved the human condition, increasing human comfort and lessening the physical demands of work, from steam power to electricity to computing. Modernization theory from the 1950s through the 1970s embodied the belief in a single route to social and economic advance through programs that attempted to bring less-developed regions into the modern world order by replicating but shortening the path to modernity. Theorists and developers later came to speak of "appropriate technology" fitted to local circumstances, as they realized how difficult it was to introduce new devices and practices into cultures quite different from those of western Europe or the United States; the term itself underscores the earlier and modern absence of understanding of the multiplicity of technologies and of the importance of local circumstances in how they are adopted and used.[16]

The question of agency is also key to an analysis of modernity, both as a program of Enlightenment rationality and as a project of emancipation. It is central to efficiency, which during modernity became a human attribute rather than a characteristic of the deity: it came to describe human agency, rather than God's. A world emancipated from tradition and superstition by human judgment, and categorized and administered by human rationality, was increasingly human built, constructed through humankind's own resources and governed by laws of its own creation. This was expressed in the concept of humanity's having "come of age," or reached "maturity," being willing to act courageously on its own judgment to fashion its own future, free of the restraints of history or naive faith.[17] Modernity as emancipation was expressed most emphatically in the Protestant Reformation and the struggles of protesting Christians for freedom from what they believed were the outmoded institutions and dogmas of the Catholic Church and which derived their power not from material necessity or human judgment but through custom or decree.

The preindustrial history of efficiency concepts bears out the shift from divine to human agency. Several plausible early sources for efficiency exist; the most obvious is comparisons of output and input in early mechanics, for which there is evidence extending back into antiquity. The history of the term itself suggests another possible source, in philosophical discussions of causation and particularly in the medieval Scholastic adoption of the Aristotelian notion of efficient cause. Still another plausible source is medieval discussions of the wisdom and simplicity of God, which overlapped with early notions of economy and household management, describing how God had created and governed the divine and cosmic household. The last one is most likely the source for efficiency's association with human power, but echoes of all three sources characterized its modern use.

An interest in output-input relationships has been traced back quite far. Archimedes asserted, as legend has it, that given a lever long enough he could "raise the world," comparing the accomplishment of a task with the resources needed to achieve it. Pseudo-Aristotle's analysis of simple machines, including the lever, suggests a similar comparison.[18] Historians have also seen an interest in efficiency in the seventeenth century, in Galileo's theory of machines, and in the early British Royal Society's interest in improving mining, transportation, and military technologies. Comparisons of output and input were made in a variety of specific instances, but the concept was not a general one, and neither Galileo nor the Royal Society used the term *efficiency*, nor did they quantify their comparisons.[19]

The word itself suggests a particular antecedent in the medieval concept of the efficient cause, based upon the ancient Aristotelian system of four causes, in which the efficient one was the active and immediate principle that produced change.[20] Thomas Aquinas was prominent among the medieval scholars who adopted Aristotle's system as part of a thirteenth-century program reconciling the pagan philosopher's work with the principles of Catholicism; the intellectual tradition of Scholasticism was founded upon this Aristotelian-Catholic synthesis. One outgrowth of this synthesis was the description of God as the efficient cause, or prime mover, a practice that endured into the seventeenth century and can be seen in Spencer's analysis of God as "the Efficient Cause of man," for he had given human form to base matter.[21]

Antecedents for efficiency conceptualized as a source of human power and authority also derive from premodern conceptions of the simplicity, economy, and power of the deity. Premodern ideas of God's nature and ways of working informed the developing idea of efficiency through doctrines of divine economy. Early ideas of economy associated with William of Ockham were also influential; according to "Ockham's razor," or the principle of economy, the simple or economical explanation of an event or phenomenon was always to be preferred.[22] *Economy* had been used to describe household management since at least 1530, but it also referred to God's managing of the cosmos, and through the eighteenth century the term denoted the "grand organization and government of life on earth" with God as "Supreme Economist" who had designed the household and kept it functioning.[23]

The modern concept of efficiency resulted from the intersection of such theories of divine simplicity, economy, and power, with premodern output-input measures, and with a theory of immediate causal agency. Efficiency of the premodern sort was neither a measurement nor a comparison. It denoted power that was sufficient or adequate, rather than a precise match between resources and task. Efficiency carried this association with power, and especially causal agency, into its modern forms.

Efficiency would thus seem to have a clear place in a human-built world. But how it works as a tool of causal agency is not clear, and formerly useful ways to account for its effectiveness are no longer tenable. Issues of cause and effect have been increasingly blurred under the reexamination of the epistemological underpinnings of human knowledge and action. It is no longer possible to describe technological matters primarily as applied science or to describe them as the products of positive knowledge of nature. Both the model of technology as applied science and, more fundamentally, of scientific knowledge as corresponding directly to the phenomena it describes have undergone serious and sustained criticism. Historical analyses have demonstrated that many technologies were developed in advance of the science that describes them or were developed by people of only limited scientific knowledge. Philosophers of the sciences have found a number of levels of interference between scientific knowledge and the phenomena it describes, the most important being unavoidable distortions through the use of language itself.[24] A discussion of efficiency is thus also a meditation on human agency, especially human intellectual agency, in a situation in which it can no longer be assumed that intellectual categories correspond to the world outside the mind. The question regarding efficiency is how, historically, people tried to make an intellectual conception into a materially effective process or tool.

The question of efficiency's place in modernity is further complicated, as are discussions of its context, by the important and sustained critique of the role of rationality in techniques of coercion and brutality. As the German thinkers Max Horkheimer and Theodor Adorno put it in their influential response to the nightmare of Nazism, *Dialectic of Enlightenment* (1947), "The totalitarian order has granted unlimited rights to calculating thought . . . Its canon is its own brutal efficiency." Critics have seen in Enlightenment rationality not only a program to ameliorate conditions of human life but also an instrument of domination, using rules of logic to exclude contradictions and reduce nuances to calculation, with the goal of rebuilding society as those who wielded it desired.[25] Efficiency, if it is characterized as a product of Enlightenment rationality and a tool for extending human power, whether as an instrument or a value in itself, will be implicated in this critique.

Three Foci

Three points of focus help to clarify the history of efficiency: attention to mechanisms of change, especially motion or change of place; techniques of agency, especially planning and discipline; and root distinctions in how efficiency was conceived and measured, between balance or growth as root concepts and between bounded measures, in which efficiency was limited by natural law, and arbitrary ones, in which efficiency was defined as a ratio between arbitrarily chosen quantities. All three may be seen in some episodes in efficiency's history, but often only one or two draw particular attention.

Things in motion are at the core of change. At its most basic, motion, or change of place, indicates that time has passed. The hands of the clock move across its face; the transitions of the cesium atom define coordinated universal time. Not only are things in motion symbolic of modernity,[26] but moving things and reshaping them is also a primary function of technology. To look for change is thus also to look for motion. Controlling the direction of change requires controlling the direction of that motion, and assessing the effectiveness of machines became simultaneously an exercise in assessing how they moved. Influential machine theorist Franz Reuleaux, longtime professor of theoretical mechanics at the Industrial Institute of Berlin, defined a machine in terms of movement: a machine was an apparatus that performed work through controlled motion.[27]

The American Frederick Winslow Taylor and his system of scientific management are the best-known historical examples of efficiency; the relationship between Taylor's famed system and his earlier work on metal cutting illustrates the underlying importance of motion. Taylor developed his management system at the turn of the twentieth century, based on work in the steel industry; the system used specialized tools, high wages, and the precise delegation of foremanship duties in efforts to consolidate management authority and stimulate productivity. Taylor had originally gained international prestige in experiments on metal cutting, in which he determined the angles and speeds of cutting blades that made them wear the longest and cut with the greatest precision; this work helped earn him a term as president of the American Society of Mechanical Engineers (ASME). Taylor's metal-cutting work influenced his theories of management and organization, especially of how things should move and in what relation to each other.

Yet this is not how Taylor's efficiency work is remembered; he is known, instead, as a management theorist, at a remove from the fundamental technological activities through which efficiency was defined. Although he was a mechanical engineer, his management theories were not fully accepted by his engineering peers, and in fact the ASME refused to publish his seminal management papers.[28] Taylor's system remained concerned with motion, however, with tools that allowed ease and speed of movement, and with the arrangement of a shop floor so that work could flow smoothly and continually.

If efficiency was part of the effort to build a modern world through human knowledge, and using human tools, then this should be apparent in the techniques through which it functioned. Two such techniques are particularly associated with the modern idea of humankind as the author of its own fate: discipline and planning. Both were practices rooted in an intellectual understanding of the human place in the cosmos and a faith in the efficacy of human action. Discipline is marked by self-control and purposeful order; planning by management and deliberate organization.

Discipline, as analyzed famously by the German sociologist Max Weber, is associated with the ascetic practices of protesting Christians during and following the Reformation and especially with developing patterns in Calvinist thought. Crucial was the concept of election: the elect were those destined for eternal salvation. Individuals, having no unambiguous way to determine if they were among the elect, sought confirmation in worldly proofs of God's favor, especially economic success through diligent work at an approved calling. Recognizing that works were not the means of attaining salvation, they nonetheless turned to them for reassurance of the efficacy of their faith.[29] The notion of efficacy was central; discipline was intended to make things effective. It could take many forms: social, personal, legal, or mechanical. It operated from above, in a coercive or normative manner, such as through laws or military commands, in which it was designed to increase the likelihood that regulations or orders were followed. It also operated from below, through social behavior, to reinforce order, security, and safety.[30] A mechanical form of discipline was embodied in the physical supports and structures intended, in a machine, to guarantee its effective action by directing and limiting how it moved.

Planning was another central feature of modernity and modern technological society and was intimately associated with discipline. It involved the surveying and apportioning of resources to achieve a desired end and thus embodied attempts to direct the course of change. Of consequence was the concept of planning itself and not merely instances of planning; critics have conceived of it as essentially technological, an extension of the techniques that had proven successful with industrial machines. Weber encapsulated planning within the terms *rationalization* and *modern rational organization;* they described a capitalist industrial order continually concerned with balancing accounts and thus organizing life so that things could be calculated and such balancing made possible. Planning looked forward and aimed to usher in an envisioned future.[31]

The third focus is on two sets of root distinctions that make it possible to differentiate important varieties among the many uses of efficiency. The first is the difference between static efficiency, allied with a notion of balance and emphasizing the conservation of measured resources, and dynamic efficiency, tending not toward equilibrium but allied with concepts of growth. This distinction is informed by the nineteenth-century developments at the foundation of efficiency's definitions in engineering: the development of the laws of thermodynamics, including the conservation of energy, and a shift in the sciences away from ideas of balance to ideas of growth. New concepts of dynamism and change came to characterize work in the physical sciences in early Victorian Britain, increasingly replacing the idea of balance, in which nature and natural or economic systems were seen as tending toward equilibrium, through changes of one sort offsetting changes of another kind.[32] Although in its formal, physical definition *efficiency* was allied with the notion of balance, in the sense that a process or machine could at best produce work equivalent to the amount of energy with which it had started, it was also used in ways corresponding to visions of change and progress.

The second root distinction is between efficiency measured arbitrarily and efficiency bounded by natural, fixed limits; it, too, grows out of the laws of thermodynamics. The difference is between quantifying efficiency as the ratio of widgets produced to the rate of ventilation, for example, and as the percentage of available energy a machine returns as work. In the first example the standard was chosen arbitrarily; in the second it was limited by the laws of energy. In establishing relationships between motion, heat, and energy and by postulating that although these quantities can be transformed they cannot be created, conservation laws provided a theoretical upper limit to the effectiveness of a machine. Arbitrary measures, in contrast, were not governed by theoretical maximums and could, and sometimes did, exceed 100 percent. But arbitrary measures could still be allied with balanced or static conceptions of efficiency. In other words, these two sets of distinctions, between static and dynamic conceptions of efficiency and between bounded and arbitrary measurements, did not map onto each other. Efficiency bounded by natural, physical laws was often allied with efficiencies conceptualized as balanced or static, but it was possible to include such a naturally bounded measurement within a broader, dynamic conception of efficiency, in which the goal was not conservation but growth. One distinction concerns how efficiency was conceived; the other concerns how it was quantified.

What follows is a multiplicity of efficiencies, chosen both for variety and seriousness. The examples range from the seemingly trivial, measures of personal leisure efficiency, for example, to the clearly momentous, as in the analysis of efficiency and slavery. Their seriousness lies in how efficiency was treated, as a matter of comparison or ratio, often measured, and part of a process of change. With two exceptions, the efficiencies here were purposeful, either deliberately sought or deliberately uncovered, and not accidental or merely rhetorical. Nor are they bound to a single time or place; many of them supported transcendent views and hopes that reached far beyond the circumstances in which they were invoked. Their multiplicity provides a view beyond any one local historical context to the larger phenomena of industrial modernity itself.

CHAPTER ONE

Static and Dynamic Efficiency

The Waterwheels of Smeaton and the Franklin Institute

The scientific and practical value of eminent engineer John Smeaton's water-wheel trials had been in dispute for nearly a century when, on December 23, 1843, a letter from Robert Mallet, member of the British Institute of Civil Engineers, appeared in the *London Mining Journal*. Mallet took exception to the Franklin Institute's estimation of the value of its recent and expensive series of experiments on the performance of waterwheels, which he claimed had "added nothing important to our knowledge"; "our great Smeaton," he argued, had done it all long before.[1] Shortly thereafter James Whitelaw wrote to the *Glasgow Practical Mechanic and Engineer's Magazine*, protesting that a recent paper had "sadly undervalued the experiments of Smeaton, and . . . attempted to give to the experiments, made by the Franklin Institute, an importance which they do not deserve."[2] The Franklin Institute responded in the pages of its own journal, "for the sake of scientific truth," as its author put it, inscribing Whitelaw "very low, indeed, upon the roll of truth" and predicting he would pass into deserved scientific oblivion.[3]

Smeaton and the researchers at the Franklin Institute were both interested in managing the power or force of water in motion, in managing a potentially turbulent flow to perform useful work, but the difference between their approaches was fundamental. Smeaton used a dynamic prototype of efficiency that led him to consider foundational questions of how matter moves in the world. The Franklin Institute, in contrast, used a static prototype that did not engage the waterwheel while it was in motion but simply measured the effect when it was done. Smeaton used his waterwheel to enter a highly abstract debate about how motion should be measured; the Franklin Institute used its to avoid theoretical nuances, testing more than one thousand configurations and publishing full results.

How Smeaton and the Franklin Institute managed the perimeter of the wheel was the crucial difference. They both set it to move at the same velocity as the water, so that it ran smoothly onto the wheel and did not splash against it, thus minimizing turbulence. Smeaton also used the velocity of the wheel's perimeter to calculate what he called "virtual head"; it was a true measure of the wheel in motion. The Franklin Institute's researchers figured instead from the before and after states; their calculations of effect did not engage the wheel while it was moving. One was a dynamic measure, the other static.

Historians have described both sets of trials as attempts to measure waterwheel efficiency,[4] and both Smeaton and the Franklin Institute did use a prototype of efficiency, measuring the waterwheel's effect in a ratio of work done to the power, or force, used. But they did not use those words and spoke instead of "used effect," "natural effective power," or "mechanical power," using terms that lacked clear and agreed-upon definitions. This was part of the reason for the hostility: beneath it lay the question of who had the right to define the terms. Efficiency itself did not yet exist as a defined and measurable concept and would not reach its mature technical form until a number of issues converged in the founding of the laws of thermodynamics and energy conservation later in the nineteenth century. But the distinction between Smeaton's approach and that of the Franklin Institute would remain fundamental: efficiency would take dynamic forms, associated with growth and transformation, and static forms, related to conservation and balance. The similarity between them would also remain fundamental: efficiency would become a method of managing turbulence.

Considerable complexity underlay both Smeaton's and the Franklin Institute's waterwheel trials. As episodes in early industrialization, they occupied a context similar in an important sense, but the aims and institutional connections of both sets of trials were markedly different. Variety in the construction and configuration of waterwheels raises another level of complexity, as does the dispute over how to measure their performance. A discussion of Smeaton's entry into this debate illustrates both the mathematical and conceptual complexity that he and Franklin Institute researchers confronted. It illustrates the deep and fundamental disagreement their measures of effect would seek to overcome.

Waterwheels in Question

Smeaton and the Franklin Institute worked in different contexts. Smeaton was the first to call himself a civil engineer, and he helped to found engineering as a profession in Britain. The experiments with waterwheels brought him into the leading scientific circles of his day, where he used them to engage in philosophical debates. The Franklin Institute was an organ of technical education and industrial improvement, devoted to technological advance by bringing the fruits of science into application rather than advancing fundamental knowledge. Smeaton was interested in the general philosophical principles behind machine performance; the Franklin Institute was interested in getting particular machines or devices to work.

The context for both studies was also a period of early industrialization and the growing use of machines to manage nature and natural processes. Larry Stewart has put it most forcefully: what made possible the increasing use of machines was an attitude that "nature can, and should, be tamed."[5] The waterwheel experiments were attempts not only to manage a natural process for use, or to manage a flow of water to perform work, but also to measure how well it had been managed. Better-managed wheels not only did more work; they, and the water in them, also moved more smoothly.

How to measure the performance of a waterwheel remained in dispute during the eighty years spanning Smeaton's and the Franklin Institute's trials. It was a practical problem: not knowing how conceptually to relate the source of a waterwheel's motion to the work it produced made it difficult to decide where and how to construct one. Builders might have to decide whether to put a wheel on a stream with little water but a high fall or on a full stream with a low fall. They were unsure how the water acted on the wheel, not knowing whether it should go under it or over the top.

Waterwheels had been important sources of power since the medieval period. By Smeaton's day they came in three basic types: overshot wheels, in which the water flowed over the top of the wheel; undershot wheels, in which the water flowed beneath the wheel; and breastshot wheels, in which the water struck the wheel somewhere in between, perhaps halfway up. Smeaton tested overshot and undershot wheels; the Franklin Institute tested all three kinds. Smeaton conducted his tests in 1752 and 1753, and in 1759 he reported the results to the Royal Society,[6] the most prestigious scientific body in Britain; his work was recognized as a contribution of exceptional merit, and he was awarded the society's coveted Copley Medal by unanimous vote of its council.[7] The Franklin Institute ran its trials from spring through December 1830 and published the results in its *Journal*.

It is difficult to appreciate the problems in conceptualizing the performance of a waterwheel in the eighteenth and early nineteenth centuries. Nearly all the terms in which a machine's performance is now measured were then in dispute, including *power* and *force*. By Smeaton's time there was one area of agreement, at least in practice: that work, or output, could be measured as the product of mass times distance, or the product of the mass of a body moved multiplied by the distance it was moved. But work was still often called other things, such as "effect." A bigger problem was how to characterize and measure motion on the front end, the input, or what made the machine go; this had been the cause of dispute for a century. Other fundamental issues also remained ambiguous; Benjamin Wilson, for example, Smeaton's correspondent and himself later a Copley Medal winner, believed electricity and gravity to be identical.[8] Terms in use during the eighteenth and nineteenth centuries to describe machine performance included *force, effect of force, mechanical power* (Smeaton's term of choice), *vis viva* (living force), and *force of a machine.*[9] Not until later in the nineteenth century were the now essential terms *energy* and *efficiency* given clear quantitative definitions. Much of the controversy took the form of an opposition between theoretical ideas and the actual practice of working mechanics, and even when the theory and practice of waterwheels were reconciled, in the work of French military engineer Jean Charles Borda, it took decades to become known.[10]

Among numerous attempts to quantify the workings of waterwheels in the eighteenth century, Smeaton's was among those most concerned with the practical utility of the device.[11] Smeaton objected to the widely known measurements advanced by the French theorist Antoine Parent and subscribed to by the popular British author and lecturer John T. Desaguliers and the distinguished Newtonian Colin MacLaurin; Parent had found the undershot wheel to be the most effective, having determined that it could produce an effect equal to $\frac{4}{27}$ of the weight of the column of water by which it was driven.[12] Smeaton believed his experiment had disproved Parent's analysis.

The debate Smeaton entered into was an important one, and highly theoretical; he nonetheless entered it as a self-described "practical man." At issue was how to determine the quantity of motion, which for him bore directly on mechanical practice; it was a question of measuring what was available to effect movement of a weight, or to effect work. He could easily tell how much work a waterwheel had done; what he could not determine was how well it had used the motion, or force or power, that made work possible. Norman Smith, in analyzing Smeaton's work, put the problem precisely: "Was the quantity of motion, or 'momentum,' of a moving body to be defined as the product of mass and velocity (MV) which was the Newtonian view; or was Leibniz's 'living force' (*vis viva*) the product MV^2, the correct measure? . . . The one measures a force acting for a given time (Ft = MV) and the other measures a force acting over a given distance (Fs = $\frac{1}{2}MV^2$)."[13] Both quantities, in Smeaton's time, were taken as measures of force. The millwright seeking to maximize the effectiveness of a waterwheel was put in a position of selecting between competing theories, one of which emphasized the time during which an effect was produced, while the other emphasized the distance over which a force or power had acted.

Smeaton considered proper conceptualization of the quantity of motion vitally important in the mechanical arts, although some involved in the debate believed their investigations had little effect on, and were of little interest to, those engaged in practical pursuits. The cyclopedist Ephraim Chambers agreed with MacLaurin that "no useful conclusion in mechanics is affected by the disputes concerning the mensuration of the force of bodies in motion."[14] Many described the dispute as a semantic one and believed the practical effect was the same however the quantity of motion was measured.

Smeaton disagreed. The thinking of philosophers and practical men should be considered distinct, he asserted, for philosophers could lead practical men astray. A proper measure of the quantity of motion and a useful understanding of the behavior of objects were indeed important to practical pursuits. He believed that the "first simple principles of science" related to "the practical and operative parts of mechanics, upon which much of the active business of mankind depends."[15] Philosophers were apt to speculate on nonexistent materials, such as perfectly hard, nonelastic bodies, which were useful entities for thinking about motion and how it was transferred from one object to another. Practicing mechanics and engineers should treat such theories with suspicion. "What ill effects can result to practical men, if philosophers should reason wrong concerning the effects of what does not exist in nature," Smeaton asked, "since the practical men can have no such materials to work upon, or misjudge of?" Because, he answered, those who inferred that hard and soft nonelastic bodies suffered an "equality of effects" could mislead the practical men "concerning the sort with which they have abundant concern, to wit, the non-elastic *soft bodies,* of which water is one, which they have much to do with in their daily practice."[16] Those dealing with water should not be misled into thinking it would behave as a hard body and into expecting and designing for effects that are impossible to achieve.

Smeaton and the Moving Wheel

John Smeaton is an engineering icon. He built the Eddystone Lighthouse on a rocky, wave-battered promontory in the English Channel; its image sits at the top of the seal of the Royal Institution of Civil Engineers. Smeaton completed the

Eddystone in 1759, the year he published his waterwheel results. The Eddystone made him famous; the waterwheel experiments established his reputation within scientific circles.[17]

Smeaton was born in 1724, son of a Leeds lawyer, and he grew up on the modest family estate of Austhorpe in Whitkirk Parish. He attended Leeds grammar school and studied law for two years in London, after which he returned to the family home to devote himself to a career as an instrument maker. Smeaton had been interested in mechanical pursuits since childhood and had found a mentor in the clock and instrument maker Henry Hindley of York, where the Smeatons had family connections. In 1748 he established his instrument-making practice in London, where he devised improvements to the air pump and developed a mariner's compass, accounts of which were published in the *Philosophical Transactions* by the Royal Society. Within four years he employed several craftsmen and counted among his clients John Ellicott, for whom he built a telescope; William Matthews, for whom he completed a precision lathe; and his close friend Benjamin Wilson, whom he had supplied with apparatus for electrical experiments since the early to mid-1840s.

In 1752 Smeaton began his experiments on waterwheel performance as part of a more general investigation of the motive force of wind and water; it was during this period that Smeaton began his career as an engineer, writing in 1753 that he intended to make "engineery" his profession.[18] He was elected to the Royal Society at the age of twenty-eight. Much of his work was in the construction of what would now be considered public works: canals, harbors, and lighthouses. He also built steam engines and other machinery. In 1771 Smeaton founded the Royal Society of Civil Engineers, which grew out of a dining club of engineers who regularly met in London to testify before Parliament about public works. He used a hierarchy of responsibility on his projects that has become familiar in engineering practice: a consulting engineer stood at the top of the hierarchy; an engineer who was resident on site during construction was next; and assistant engineers were hired as needed. Rather than receive a wage, Smeaton charged consulting fees; near the end of his career he admitted to charging particularly exorbitant rates as a way to protect his time.[19]

Although Smeaton completed the trials in 1753, he did not report on the experiments until 1759, at which time he was engaged in rebuilding the Eddystone Lighthouse. During the intervening years Smeaton had made a thorough study of Continental work on engineering and machines, paying particular attention to Bernard Forest de Belidor's *Architecture hydraulique* (1737–53), and had journeyed through the Low Countries inspecting examples of engineering practice. He had also contributed papers to the Royal Society on lightning damage to structures and on weather patterns. A. W. Skempton's list of Smeaton's engineering works includes the design and construction of three major waterwheels during this period, from Smeaton's own lifetime list of the forty-six he had completed.

Smeaton's Experimental Device

Smeaton's waterwheel experiments required careful choreography. His model stood about chest-high, with a wheel that was twenty inches in diameter. It used a vertical cistern that had a hand pump with a horsehead, or pump jack, designed to keep the cistern full during an experiment by raising, from the reservoir below, water that had already passed over the wheel. During a trial the pump was operated with a steady rhythm and consistent length of stroke, to keep the level in the cistern and the head of water even and to keep track of the amount of water expended. An aperture could be adjusted to make the wheel overshot, undershot, or breastshot; it was opened at the start of a trial. Attached to the axle of the wheel was a rope threaded through a pulley anchored about fifteen feet above the model, supporting a pan; when the rope was engaged, the pan raised a weight. During a trial, when the wheel was running smoothly, the rope was engaged and, at the end of the trial, disengaged. The dynamism of Smeaton's method lay in operating the pump and engaging the rope, not merely because they were active motions themselves but because they allowed him to engage with the wheel while it was in motion. They allowed him to calculate what he called "virtual head."

When engaged, the rope wound around the axle and raised the weight. Smeaton then measured how high the weight had been raised, multiplied that amount by the weight, and compared the product with the quantity of water that had descended the measurable perpendicular distance in activating the wheel. This comparison would define what Smeaton called the "natural effective power" of water as a prime mover.[20] Smeaton defined "the word *Power,* as used in practical mechanicks . . . to signify the exertion of strength, gravitation, impulse, or pressure, so as to produce motion," and argued that "no effect is properly mechanical, but what requires such a kind of power to produce it."[21]

The Dynamic Measurement of Virtual Head

The key to Smeaton's analysis of waterwheels was his calculation of what he termed "virtual head." Determining virtual head required him to confront the waterwheel while it was in motion, rather than simply measuring its state before

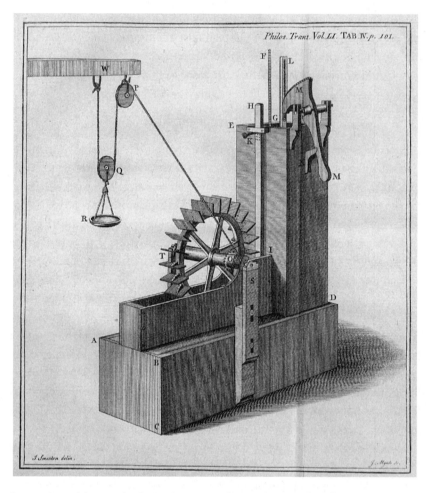

Philos. Trans. Vol. LI. TAB. IV. p. 101.

Smeaton's model waterwheel. "An Experimental Enquiry concerning the Natural Powers of Water and Wind to Turn Mills, and Other Machines, Depending on a Circular Motion," *Philosophical Transactions of the Royal Society* 51 (1759): 101, table 4. Courtesy of the Wangensteen Historical Library of Biology and Medicine, University of Minnesota.

and after a trial. This quantity was a dynamic measurement and led him into philosophical debates about how to characterize matter in motion. The problem with motion in machines was that it caused friction, and it was in order to manage friction that Smeaton introduced virtual, or effective, head. He measured friction by using a counterweight to set the wheel in motion at a velocity equal to the velocity of the wheel when the water was moving it, whether or not the water was in flow. In other words, he found a velocity for the perimeter of the wheel equal to the velocity of the water, one in which opening the aperture and letting the water run neither sped up the wheel nor slowed it down. Smeaton used this velocity to calculate what he termed the virtual rather than the actual head.

The key to Smeaton's calculation of virtual head was the velocity of the blades of the wheel at its perimeter. It was a dynamic measure, taken while the wheel was in motion. Smeaton's measure of virtual head was unique, and although he used it to manage the effect of friction, it led him into more fundamental discussions of how to characterize and measure motion. Measuring the virtual head involved four steps. First, the velocity of the wheel was determined by use of a counterweight, as just described, so that its velocity was the same whether the water struck it or not. The velocity of the water was thus equal to the velocity of the wheel. Second, using what he called the "common maxims of hydrostatics," Smeaton used the water's velocity to calculate the "virtual or effective head."[22] Third, he calculated the volume of water expended by the number of pumps of the handle needed to keep the reservoir full, having already determined the volume of water raised by each full stroke of the pump handle; the volume of water was then multiplied by the unit weight of water to yield the total weight of water expended. Fourth, and finally, the weight of water expended was multiplied by the virtual head to yield what Smeaton termed the "power" available to produce mechanical effects. The key was the measurement of the velocity of the water, a measurement taken while the wheel was in motion.

Smeaton concluded that undershot wheels had a higher ratio of effect produced to power expended than theorized: $\frac{1}{3}$ rather than the $\frac{4}{27}$ predicted by Parent and others. He also found that overshot wheels were, on average, twice as effective as undershot. His table of results for undershot wheels included thirteen categories of data, such as the height of water in the cistern, turns of the loaded waterwheel, the weight of the water expended, and maximum load. No units appeared in the columns reporting Smeaton's calculations of power, effect, and their ratio; they were dimensionless numbers. The units that had disappeared had canceled each other out in the course of his calculation. This was a prototype of efficiency, a ratio of output to input, but not yet named as a general quantity.

Smeaton's choice to use virtual head in the trials, rather than total fall, suggests that his interest was broader than immediate practical utility. His work challenged the authority of Cambridge natural philosophers such as George Atwood and Isaac Milner and their use of experimental demonstration devices. In calculating virtual head, Smeaton entered an ongoing philosophical and religious dispute over the intrinsic properties of matter and the causes of motion.[23] His calculation employed a measure of available mechanic power equivalent to Leibniz's description of *vis viva*, or living force, which many English natural philosophers had denounced as a materialistic principle synonymous with innate activity and perpetual motion.[24]

Smeaton described a highly technical and abstract debate in terms of his machine. He related mechanical effect, or work, to both the raising of a weight in the pan and the lowering of the weight of water from the height of its head. He made clear that what mattered was not how much time elapsed in performing work but the distances involved. He did not vary the time of his experiments; all of the trials ran one minute. He did vary the height of head in the cistern, but his essential calculation turned instead on figuring virtual head.

Contributing to Theories of Matter in Motion

Mechanical power, as Smeaton described it, was a matter of continuing motion, subject also to the continued action of friction. Smeaton's use of a dynamic measurement thus enabled his Cambridge detractors to dismiss suggestions that his work had implications for the proper understanding of Newtonian mechanics while still accepting its potential practical value. Static structures such as bridges could be analyzed fruitfully in terms of Newtonian mechanics; bodies in motion posed a different sort of problem and could not be accepted as tests of theory. "Friction and other resistances," George Atwood wrote in 1784, caused "the motion of bodies, which are the objects of experiments, to differ from the theory." Simon Schaffer has argued that Atwood may have developed his own famous machine in response to Smeaton's experimental techniques, especially in the paper of 1776 on mechanical power and velocity.

The 1776 paper was one of two more on the foundations of matter in motion that Smeaton contributed to the Royal Society in which he argued for "mechanical power" as the common measure by which concepts such as force, momentum, or the quantity of motion should be defined. He also continued to contribute papers and notes on more practical matters. Neither of the two additional papers on mechanics used a model of a device in practical use; instead, Smeaton built apparatuses specifically to test matter and motion theories.[25] One used a horizontally rotating arm propelled by a weight attached to its axle; it was a wheel without the water. The other was an ingenious ratcheted device designed to absorb forces and then spring back when the ratchets were released, simulating bodies in collision. It was like water without the wheel.

With the three papers on mechanics Smeaton stepped onto ground that was contested both socially and philosophically. Not only were concepts such as momentum and power in dispute but the question of who had the authority to define them was unsettled as well. Natural philosophers in eighteenth-century Britain increasingly needed to demonstrate the utility of their knowledge in order to gain public authority. The rhetoric of power, of the ability to control nature to affect public improvements, merged with the rhetoric of experiment;[26] Smeaton's experiments drew upon the authority of both traditions, the practically effective mechanic and the natural philosopher as experimentalist. Smeaton's use of a dynamic measure in his assessment of "mechanical power" made his results of general theoretical interest and, by raising the problem of friction, gave Cambridge mathematicians an opening to dismiss the theoretical implications of his work and thereby defend a distinction between analytical and mathematical rational mechanics, and practical or industrial mechanics.[27]

Smeaton's papers on mechanics were reprinted and distributed widely. Not only did his work supply a basic principle of waterwheel design—when possible, use an overshot wheel; it also led him into fundamental questions about the nature of motion and the structure and behavior of matter. *Virtual head* defined available input and thus gave Smeaton a commensurate quantity, mechanical power, to compare with the work the wheel did as output. In using the term, Smeaton confronted the actual motion of the machine.

Continuing Practical Confusion

John Farey observed early in the nineteenth century that Smeaton's improvements in waterwheel design had slowed the spread of steam engines, having shown how "the performance of almost all the mills then existing, could be doubled with the same supplies of water."[28] But Smeaton's waterwheel trials, despite having yielded general guidelines for wheel configuration, remained insufficient in practice, as illustrated in continuing practical confusion.

Practical men were indeed interested in these questions, as a brief survey of correspondence on this issue between members of the American Philosophical Institute and the Franklin Institute reveals. William Waring argued in 1790 that the theory of mills had been of little use in practical mechanics because it was marked by several errors. Of particular interest to him was the proper relationship between a definition of "available force" (a term he did not define but nevertheless used to speak of the stuff that would make the waterwheel run) and the velocity of water against the wheel. Atwood and others had erroneously supposed "the momentum or force of the water against the wheel, to be in the *duplicate ratio of the relative velocity.*" Instead, Waring suggested that the force of a stream of water against a wheel was in "simple direct proportion" to the velocity of the wheel relative to the velocity of the water. Using this principle would bring "the theory to agree remarkably with the notable experiments of the ingenious Smeaton," and would yield the same ratio of wheel to water velocity that Smeaton had found experimentally.[29] Three years later Waring further explained his corrective to mill theory by clearly distinguishing between the force of water impinging against a mill wheel and the effect produced by that impact, or "impetus."[30]

Waring's use of terms—and his need to contribute an additional paper solely to distinguish between *force, momentum,* and *effect*—illustrates the continuing uncertainty over the meanings of terms in both practical and theoretical use. He used the terms *force* and *momentum* interchangeably to refer to what he called in his addendum the "impetus" of the water against the wheel. The "effect" of the wheel, the measure of which would contain a squaring of velocity, was itself carefully distinguished from *impetus* and thus from *momentum* or *force.* Waring's attention to so basic a term as *velocity,* and to its proper function in calculations of effect, underscores the practical confusion over fundamental mechanical concepts.

Such practical concern was also expressed in Britain. William Wollaston, in the Royal Society's 1805 Bakerian Lecture, used the driving of piles as an example of the practical consequences of the dispute over how to measure the quantity of motion: measuring "through a determined space is of greater practical utility, as it occurs in the usual occupations of men; since any quantity of work performed is always appreciated by the extent of effect resulting from this assertion."[31] The pile-driving example appeared also in the *Journal of the Franklin Institute,* in which an anonymous author argued that the proper measure of "powers" was as the square of their velocities; if power were measured by simple velocity, it condemned "as ignorant so great a number of men eminent for practical mechanical knowledge" who counseled that pile-driving weights be lifted to the highest point convenient before being released.[32] Signing only with initials, J.P.E., who also wrote to the institute on the "force of bodies in motion," was soundly criticized for confusing terms such as *momentum* and *power,* a gentleman from Jersey writing that he had left "no room to doubt his meaning of the word momentum, as being synonymous with power; and, consequently, that a double velocity of water gives a double power, which leaves him in the dilemma, of having endeavored to support that a double power of water will produce a quadruple effect, as he says a double velocity is known to do."[33] Confusion continued.

The measure of friction and its status as a force, or power, was also in dispute.

Smeaton had employed two methods of accounting for friction in his waterwheel experiments: use of a counterweight to generate an equal velocity in the wheel as a measure of its friction; and calculation of virtual, or effective, head to account for friction experienced by the water en route to the wheel. Neither of these methods was available to the millwright engaged in design and construction. Unless the velocity of the wheel could be determined—a practical, empirical deduction—the velocity of the water could not be found; absent a measure of the water's velocity, the effective head could not be found. Smeaton had observed that "the *virtual head* bears no certain proportion to the *head of water*."[34] Another anonymous correspondent in the *Journal* reported on his attempts to determine whether friction should be measured according to the duration of its effect, or according to the space over which it acted, the very problem at issue in the *vis viva* controversy.[35] Despite Smeaton's attempt to give practical guidelines for determining the effectiveness of various waterwheel configurations, basic terms and concepts remained muddled.

It may be that Americans simply did not keep up with theoretical and experimental developments abroad. Borda's equations for waterwheel output may have provided a sound practical basis for waterwheel practice before the turn of the nineteenth century, had they been adopted, but Borda's analysis was not widely circulated until after 1840.[36] In particular, the Franklin Institute's *Journal* may not be a reliable indicator of the state of theoretical knowledge, as theoretical topics settled on the Continent continued to be disputed in the *Journal* years later.[37]

Confusion may also have continued because those most practically affected were not addressed by the persistent argument over how to conceptualize motion and power, or force. Millwork in America was little influenced by conceptual discussions. Unlike engineers, millwrights were not familiar with many practical and theoretical developments, for they were trained largely in an oral apprenticeship tradition. Although Oliver Evans had published Smeaton's waterwheels paper in his popular *Young Mill-Wright and Miller's Guide* (1860), quantitative concepts of practical and theoretical effect remained confused. Millwrights, taught to reason qualitatively, often considered analysis "foolish and fruitless" and pursued technical advance through empirical testing.[38]

Whether genuine confusion remained over fundamental concepts in measuring effect or power or whether American millwrights simply could not conceptualize a practically adequate measure to determine the effect of a waterwheel before construction, Smeaton's analysis was widely considered practically insufficient. Those working with machines did not agree on how power, force, and measures of the quantity of motion were related to their endeavors; recognizing the practical insufficiency of existing conceptual and experimental traditions, the Franklin Institute committed itself in 1829 to making its own series of experiments on the effectiveness of water as a moving power.

Avoiding Motion at the Franklin Institute

The Franklin Institute of the State of Pennsylvania for the Promotion of the Mechanic Arts was founded in Philadelphia in 1824. Its originators were Philadelphia industrialists, improvers, and utopian thinkers seeking to offer the fruits of knowledge to the working people of the city, men such as Samuel Vaughan Merrick, iron founder, and William H. Keating, University of Pennsylvania professor of chemistry and natural science. Its goals were similar to mechanics' institutes founded in Birmingham, London, and Glasgow, site of the famed Andersonian Lectures, but it also expended significant effort on evaluating new devices and testing industrial equipment.[39]

The Franklin Institute was founded just as the Philadelphia machine-making industry entered a period of tremendous growth, as the number of practicing mechanics and engineers increased nearly ten times between 1831 and 1853.[40] Philadelphia mechanics formed a tight community, visiting each other's shops and marrying into each other's families.[41] The Franklin Institute's Committee on Science and the Arts examined and commented upon models of machines and devices for area mechanics, often as an aid in applications for a patent. Thomas P. Jones, a member of the Franklin Institute, had in fact served as head of the Patent Office, but in practice patents were granted with an "easy liberality" and proved of little help in discerning the usefulness of new machines. The Franklin Institute's process was more exacting; it was confidential, and the devices were not put on public display, but the results of its investigations were published in its *Journal.*[42]

The institute was unique in undertaking large-scale research projects into complex and difficult technological matters, of which the waterwheel experiments were the first to be concluded. In addition to its investigation of waterwheels, from 1830 to 1835 it undertook a study of boiler explosions. The waterwheel experiments, conducted from spring through December 1830, were underwritten by subscriptions from institute members and similar organizations (the cost was estimated at two thousand dollars) and overseen by a committee dedicated to the study of water as a prime mover. The first report was published in the institute's *Journal* in March 1831 and republished in London's *Mechanics' Magazine*, copies of which were imported for sale in the United States by a Boston bookseller, undercutting the institute's hope to recoup some of the expenses of the trial by selling its report. George Rennie, an authority on waterwheel performance, "having made a great many experiments on the maximum effect of water wheels" himself, told the British Association for the Advancement of Science in 1832 that the Franklin Institute experiments, "made on a more magnificent scale . . . eclipse everything that has yet been effected on this subject."[43]

An Unambiguous Measure

The Franklin Institute was interested in the relationship between power and effect, as Smeaton had been, and it too defined these concepts in terms of the components of its machine. *Effect* described a mass moved and the vertical distance it moved; *power* described almost the same thing—another mass moved (water) over another vertical distance (fall). But whereas Smeaton's wheel trials had led him to consider the theoretical ambiguities inherent in a measure of power or effect, the Franklin Institute's waterwheel experiments were specifically designed to sidestep conceptual ambiguity. The institute's study was undertaken because water's effects on different wheels had "never been fix'd by actual experiment on a scale of sufficient magnitude to settle principles upon which it is to be calculable."[44]

The institute's report described how "the different theories advanced are at variance with each other, and with practice, so that the candid theorist confesses that the circumstances, attending the action, are of so complicated a nature as to baffle his powers of investigation."[45] Smeaton's investigations, although prosecuted "with great skill and resource," had not been numerous enough to make the results of his model tests useful, and readers of the *Journal* continued to submit queries about the best construction of a waterwheel under basic fall conditions. In response, the Committee on Water Power deliberately chose the most unambiguous measures of power and effect available: the simple product of the distance of fall and weight of water expended to denote power; the product of weight raised and distance raised to denote effect. This made the institute's as much an example of early industrial testing and quality control as an experimental exercise and was in keeping with an institution that viewed itself as democratic rather than expert.[46]

The experimental apparatus approached full size. It contained a forebay of variable capacity partially surrounding the wheel, allowing water to be forced onto the wheel from overshot, breastshot, or undershot positions. The diameter of the wheels varied; the first tests used one measuring 20 feet. A reservoir to collect and weigh the water used was constructed below the wheel, having a capacity of 30,000 pounds. Attached to the shaft of the wheel was a chain, leading over a raised pulley to a bucket in which weights were placed. The committee had adopted as its measure of effect "the weight raised, and the height through which it was raised."[47] Two marks were placed on the chain; one, when it reached a certain indicator, meant that the experiment was to begin, while the other signaled the trial's end. Three assistants attended each trial, one to maintain a constant water level in the forebay, thus ensuring a constant head, the second on a platform at the hub of the wheel to operate the collection reservoir and to observe the movement of the chain. The second person rang a bell indicating the start and finish of each trial, at which time the forebay operator opened or closed the aperture that let water onto the wheel. A third person noted the time elapsed between bells and recorded general observations about the conditions of the experiment. Trials were also made to determine the weight needed to overcome friction of the machine, and allowances were made for it in computing the effect. To overcome inertia the designers employed a system similar to Smeaton's: the wheel was allowed to settle into a uniform velocity before the timed trial began.[48]

The apparatus differed from Smeaton's in several obvious ways: it was bigger and allowed a greater variety of possible configurations, such as different aperture sizes and various heights of breastshot application. But the purpose of the machine was the same as it had been for Smeaton: to test the effect of water as a prime mover. The institute employed the same measure of effect as Smeaton had: the product of weight (mass) raised by the distance it was raised.

Unambiguous Results

The institute recorded 1,381 waterwheel experiments. The results were published in tables of up to eighteen columns, including measurements of head, width of aperture, weight raised, friction, time elapsed, and wheel velocity. On each table special note was made of the most effective configurations, their ratios of effect appearing in a specially designated column. Most configurations went through 20 trials. The trials demonstrated a significantly higher ratio of effect to power expended for overshot waterwheels than had Smeaton's experiments: one trial demonstrated an effect of .872, having used 3,100 pounds of water to raise a weight of 1,287 pounds over a distance of 41.5 feet, while overcoming 65.11 pounds of friction.[49] Millwrights and engineers could search through the tables to examine waterwheel performance in different configurations. Results of all the variations tried were published; the tables were sets of empirical data and were not organized into general conclusions. Bruce Sinclair described them as an early form of engineering handbook.[50]

The Franklin Institute experiments did not investigate the dynamic relationship between fall and power or effect, which was central to Smeaton's report. The Franklin Institute did not calculate virtual head but instead reported static head and fall. In other words, it measured the wheel's performance on the basis of data taken before and after the trial, not during it.[51] Water expended was collected in a reservoir below the wheel and weighed after each trial ended; this was a static measurement. Fall represented the height, or distance between the top of the static head of water and the bottom of the wheel, which just skimmed the bottom of the tailrace; it was determined before a trial began. These two measurements multiplied together gave the amount of power. The Franklin Institute relied on measurements taken after motion had ceased, or which were not themselves affected by the motion. Smeaton, in contrast, had adopted a dynamic principle in determining head—his virtual, or effective, head—that rested on a measurement taken while the water was in motion.

The Franklin Institute conceptualized power and effect in terms of its machine and not in an abstract or general sense. Various components were measured before and after each trial and relationships drawn among them. In this way the institute's experiments were much like Smeaton's, and he had indeed used the same measure of effect, an essentially static one. But he had explicitly considered the relationship of effect to a dynamic measure, and here lay the difference between these two episodes. By using a dynamic measure in calculating virtual head, Smeaton had confronted motion; the Franklin Institute, relying on static measurements, confronted only its effects.

Smeaton used his waterwheel experiments to enter a philosophical debate over how to measure the effectiveness of a machine; a century later the Franklin Institute deliberately avoided philosophical complexity by employing two accepted and unambiguous measures of effectiveness: weight and distance. In trying to quantify the "natural effective power" of a waterwheel in motion, Smeaton had approached the problem as one of dynamics; the Franklin Institute, looking for the result, or "effect," of waterwheel motion, approached the problem as one of statics. Smeaton's results were general although widely cited by British engineers, and they remained of primarily theoretical interest. The Franklin Institute produced a handbook of more than a thousand possible configurations, requiring no intervening theory to decipher.

For Smeaton the waterwheel had provided an entry into difficult theoretical questions; it was used by the Franklin Institute to take practical action in the face of conceptual ambiguity. Both used a prototype of efficiency, but only Smeaton, in his theory of "mechanical power," had suggested it as a general concept. The measures of effectiveness or mechanical power required careful control of the motions of the machine in both cases and the elimination of unwanted influences. In setting the velocity of the perimeter of the wheel to match the velocity of the water, they both used a similar method to minimize turbulence. But Smeaton, through his analysis of virtual head, had made a dynamic measure, whereas the Franklin Institute was comparing two static states. The Franklin Institute only wanted the waterwheel to do its job; Smeaton had wanted it to reveal the secrets of matter in motion.

The complexity inherent in conceptions of machine performance, or effect, becomes apparent in the different approaches of Smeaton and Franklin Institute researchers. Beneath even the relatively unambiguous measurements made by the Franklin Institute was a series of decisions about how the world does and should behave—more specifically, about how matter, and engineers and mechanics, should behave. How explicitly these issues were addressed differed. Smeaton saw in measurements of effect a way to interrogate philosophical assumptions about matter and social assumptions about to whom such philosophical questions would matter. Franklin Institute researchers, more concerned with immediate practical utility, said little about the beliefs this expressed.

CHAPTER TWO

The Effects of Control

Gérard-Joseph Christian and Perfected Machines

César Nicolas Leblanc and his team of draftsmen spent months in 1828 examining and preparing to draw the machines at a new cotton-spinning mill near Compiègne, north of Paris. The mill was of English design, and its owners had been forgiven import duties if they allowed a full description to be published. Leblanc was principal engraver at the Conservatoire des Arts et Métiers, the central depository of industrial knowledge in Restoration France. Producing the engravings was arduous work. Leblanc and his team looked at each machine up close—at its bobbins, flyers, and wheels, its bolts and nails—then they stepped back and made sketches, moving up close again to touch it and see if their drawings did it justice. They pulled at pieces of it to see how firmly they held and slid their hands along its contours and hidden workings to get a physical sense of how it fit together. From part to part they moved and then back again, from spindle to spindle and from frame to belt, stepping away and returning until they knew the machine not only by eye but by hand and mind. Only when they knew it and how it worked, how the cotton traveled from roving to fine-spun thread, when they knew what would pull the fibers and what would twist or wind them, and when they could follow the route with their hands, only then were they ready truly to draw.

Drawing this way was active and mentally demanding. Gérard-Joseph Christian, director of the conservatory, believed that the same combination of varied physical activities and continuing mental attention was vital not only to draftsmen but also to the workers who would operate the machinery of a newly industrializing society. The English spinning machinery illustrated Christian's philosophy of industrial mechanics: it was a mechanically sophisticated system, its motions well controlled, and operating it was not mind-numbing but required intelligence and close attention. In its economy of force and materials and its good production it illustrated a form of mechanical perfection, and in its demand of mindfulness in operation it illustrated machine economy in the service of society, neither eliminating workers' jobs and livelihoods for the sake of profit nor demanding that they expend their energy in repetitive and mindlessly rote gestures.

Christian's work illustrates an idea of effectiveness that combined both static and dynamic elements in a defense of emerging industrial society, developed in a time of machine-breaking riots and precarious political stability. He was named conservatory director in 1816, as France recovered from more than twenty-five years of revolution and war, a financially and morally devastating occupation by the allied powers that had helped to defeat Napoléon, and a large war debt of its own, alongside a seven million–franc indemnity assessed in the Second Treaty of Paris. The dynamism of Christian's theory lay in a view of the transformative power of the most effective machinery, which promised to free humans from rote and exhausting labor and to create new work that would stretch workers' intellects and allow them to move their bodies in varied and far-reaching ways. Its static dimension lay in a contrasting treatment of motion: the most effective machines were the most immobile, acting only through predetermined movements of limited variety, thus conserving motive force and displaying economy in production. He used an idea of the perfected machine and a prototype of efficiency to distinguish between the movements of workers and machines.

Christian worked in the intellectual and mechanical tradition from which the formal concept of efficiency would emerge, a tradition concerned with assessing the performance of machines and which recognized the importance of motion control in achieving their greatest effects. At the mechanical level Christian's analysis was similar to both Smeaton's and that of the Franklin Institute's researchers, but he held a much broader, transformative vision of the role of effective machines in society. In the end his concept of the perfected machine had to do not only with a ratio of the work it did to the power or force it had consumed, although that was part of it; it had also to do with the social role of the machine. Effectiveness, for Christian, was not limited to the quantifiable.

Christian's Platform: The Conservatoire des Arts et Métiers

Christian's career illustrates the use of concepts of effectiveness in envisioning how to manage industrial change in a period of social and political turbulence.[1] His years at the conservatory spanned two multiyear episodes of anti-machinery protests and riots, periods of crop failure and widespread hunger, including hunger riots from the spring of 1816 through the summer of 1817, the assassination of the presumed heir to the throne and an ensuing political crisis in 1820, and repeated confrontations between supporters of liberal and traditional monarchy. A synopsis of alternations in authority between emperor and king will illustrate uncertainties at even the highest levels. Napoléon had abdicated in the spring of 1814, and the Bourbon family had been restored to the throne, in the person of the comte d'Artois, who became Louis XVIII. But Napoléon escaped from the island of Elba the following spring and marched on Paris, gathering soldiers and followers on the way, and the king abandoned the city the night before Napoléon arrived. In the wake of his defeat at Waterloo, in June 1815, Napoléon abdicated again, and the king returned to Paris a few weeks later. The restored Bourbon monarchy was in a precarious position, hampered by debt, weakened by its flight at the reappearance of Napoléon, and trapped between paternalistic sympathy for citizens and subjects and an economic commitment to the principles of laissez-faire.

Christian spoke from the platform of the Conservatoire des Arts et Métiers, an institution founded by the state to regulate industrial knowledge and promote forms of industry consistent with its vision of society.[2] While by Christian's time its foundational goals no longer governed the institution, they remained influential in ways that were apparent in Christian's work.[3] He sought to perfect French industry, the task with which the conservatory had been charged, but he defined *perfection* as more than mere improvement. To Christian the most perfect machine produced the most, the most rapidly, and with the greatest economy; it also contributed to the intellectual and physical well-being of the worker. Christian believed this would all happen as a matter of course; his was not a program of reform.

The Conservatoire des Arts et Métiers was the central depot of industrial knowledge in France and an important site for its dissemination. Its tools were various publications, a technical museum and small drawing school, and, after 1819, a program of popular public lectures. The Convention Nationale had founded the institution in 1794 as a technical museum, opening to public view a variety of mechanisms, models, and drawings previously held privately or otherwise inaccessible. It combined the patent models of the Académie des Sciences, which had before the Revolution been in charge of granting patents; the private drawing, model, and machine collection of the inventor Jacques de Vaucanson, which Louis XVI had inherited upon Vaucanson's death in 1782 and which in the 1790s had been significantly enhanced by the acquisition of models and machines from

England and the Low Countries; and machines and models assembled, some confiscated, by the Section Mécanique of the revolutionary government's Commission Temporaire des Arts and the Comité d'Instruction Publique.[4] A team of demonstrators, led by Claude-Pierre Molard, was available to explain the displays, and a draftsman was engaged to make drawings of all the machines in the collection, to be displayed alongside them and to aid in the creation of a standardized and uniform mechanical practice. The conservatory was to instruct the public in "the construction of the most perfected tools and machines, their operation, the distribution of movement, the use of force."[5]

Having the conservatory as his platform gave Christian's work prominence, and he used it to advance a more theoretical orientation for technology, based on the unique dimensions of industrial mechanics. Industrial mechanics treated real bodies moving over time, in contrast to rational mechanics, Christian argued, which instead treated ideal bodies in ideal circumstances, such as point masses at an instant. Machines were real and substantial, and their mass did not cohere at a single point. Christian insisted upon the novel character of industrial mechanics throughout his career. He had aspired to a new chair of technology at the prestigious École Polytechnique, and the conservatory was a second choice, though it may have better fit his goals. Real industrial operations, in the crafts and manufactures, were the conservatory's province.

The institution's foundational task was immense: the improvement of the arts of French industry in the service of liberty, allowing it to shake off the yoke of foreign industry and become independent.[6] Worries about industrial stagnation had not been assuaged in the years since the conservatory's founding. In the introduction to his study of the Compiègne spinning mill Leblanc himself enumerated a variety of causes for France's relative decline, stemming from revolution and war. The demand for industries manufacturing anything other than arms or military equipment had withered, and large numbers of the population had taken up arms, interrupting both production and consumption of industrial goods. England, in contrast, had during a period of relative domestic peace developed a mode of spinning of which the French were completely ignorant. French inferiority was a frequent theme in reports by travelers who had visited England, and Leblanc cautioned that France must do whatever was in its power to achieve industrial parity. Charles Dupin, who later became a lecturer at the Conservatoire des Arts et Métiers, was one of many who undertook a trip to Britain to examine the state of its industry and education.[7]

Tensions between scientific and mechanical knowledge were evident in Christian's career, where they occasionally took the form of personal antagonisms.

The intellectual and scientific circles in which Christian moved were pre-professional, still tied to patronage and personal authority, and personal dislikes played important roles.[8] He regularly associated with scientific luminaries, including André-Marie Ampère, Joseph-Louis Gay-Lussac, and François Arago, and he had served in the ministry of the interior, inspecting manufactures, during the brief administration of Lazare Carnot. But he thought the work on mechanics by his peers was too theoretical for use by those engaged in making and operating machinery and spent much of his career attempting to build a bridge between theory and practice. Before coming to the conservatory, he had held a chair in physics and experimental chemistry at the École Centrale de Namur and then a professorship of mathematics at the Lycée of Brussels. The work of the German Johannes Beckmann influenced him greatly; Beckmann had pioneered the term *technology* and conceived of his subject in administrative and economic terms. Christian proposed a chair of technology at the École Polytechnique in 1816 but was rebuffed in ridicule, called nonobjective and pretentious; the mathematician Augustin Cauchy, student of the mathematicians Joseph-Louis Lagrange and Pierre-Simon Laplace, was instead promoted from instructor to professor and given charge of courses newly titled "industrial mechanics." Although the conservatory was not his first choice, Christian embraced his position with great hopes. The appointment was an affirmation of his work; it entailed rejection of the candidacy of the son of Claude-Pierre Molard, the previous director, and thus evinced a desire to give the institution a new direction. Molard *fils* was appointed Christian's assistant director. Christian resigned from the conservatory in 1831 and died of cholera the following year.[9]

The Conservatoire des Arts et Métiers had only an indirect teaching function. Neither systematic nor comprehensive preparation for a vocational or engineering career was provided, with two exceptions: a short-lived school of cotton spinning and an important drawing academy. A shortage of mechanics able to oversee the operations of new textile machinery had led to the opening of a school on spinning machinery in 1808, which operated until low enrollments forced it to close in 1814.[10] In 1806 the conservatory had begun to offer instruction in technical drawing to young men in their teens, often the sons of artisans, in what came to be called the Petite école; its professors, Leblanc and his successor Jacques Eugène Armengaud, became widely influential in developing new standards of draftsmanship. The most remarked-upon of the conservatory's formal teaching endeavors was its series of open public lectures instituted in 1819, given by noted savants and attended by adults, including workers, mechanics, and industrialists.[11]

Other institutions of technical education founded in the same period did have direct teaching roles, among them the elite École Polytechnique, established in 1794 to provide mathematical and scientific training for engineers destined primarily for state service, and the prototypes of the Écoles des Arts et Métiers, which would become the training ground for the engineers who would serve French industry. The École Centrale des Arts et Manufactures was founded near the end of Christian's career, in 1829, and became an important institution of training for civil engineers.[12] The conservatory, in contrast, offered no formal schedule of classes and had no program to award certificates or otherwise mark the completion of a course of study. It was a success as a museum, and its open lectures were popular, in part because the speakers were prominent: Jean-Baptiste Say on industrial economy, Charles Dupin on mechanics, and Clément-Desormes in chemistry. Their audiences numbered from fifty to two thousand; among them was Sadi Carnot, the son of Lazare Carnot, whose theory of the motive power of heat would form one of the foundations of the science of thermodynamics.[13]

The effects of a history of political instability on mechanical education can be seen in the early fortunes of the lecture program. Early in 1820 the duc de Berry, heir presumptive to the French throne, was assassinated outside the Paris Opera, and the reaction ushered in a period of more conservative government hostile to popular education. Almost immediately a new government had been formed under the duc de Richelieu, who, although not particularly conservative, was less liberal than his predecessor Élie Decazes; within two years Richelieu's government itself gave way to the ultraroyalist ministry of the comte de Villèle. Villèle dismissed the liberal philanthropist the duc de La Rochefoucauld from the governing board of the conservatory and oversaw the closure of a number of schools, including for four months the École de Médecine (in the winter of 1822–23), and the suspension from their posts of a number of professors of liberal leanings.[14] The police began surveillance of the conservatory's public lectures, noting that both the speakers and audience expressed hostility to the restored monarchy and the clergy.[15] Parisian mechanics had a history of agitation: angered at the long stranglehold on patents by the Académie des Sciences, they had during the Revolution used intimidation to ensure adoption of a new patent system, congregating outside the Convention Nationale as it deliberated on the matter.[16] By Christian's time the conservatory was well settled on the old site of the priory of St. Martin of the Fields, in the center of Paris's district of mechanics' workshops and small manufactures.

Christian was barely involved in the teaching initiatives of his institution, despite their prominence and their having been founded during his tenure as director; in fact, he had always been more interested in philosophical than pedagogical matters. He turned to exhibitions and publications as a way of disseminating his analysis, serving on the juries of exhibitions and mounting others himself; analyzing and cataloguing the conservatory's collections; writing and publishing several important texts on industrial mechanics, including a serial that ran for four years; and collaborating with Leblanc in using mechanical drawings and engravings to distribute news of mechanical practices and developments. Christian's work was primarily concerned with the demonstration of machines, whether in their full materiality in exhibitions and the museum or in texts and engravings, and his endeavors were united by an interest in identifying and publicizing the best machines and processes, those that performed well and with good economy. His, and the conservatory's, was an intermediate approach to mechanical education, lacking the immediacy of traditional forms of apprenticeship and not yet training mechanics or engineers in applied science.[17]

Exhibitions had become a recognized form of industrial education during the years of revolution and empire. Industrial exhibitions showcased the best of French products and helped disseminate news of advances in processes and machines. The first was held in revolutionary year VI, others in years IX and X, and another under the empire in 1806. Christian collaborated with the interior ministry and members of the conservatory's governing board in mounting an exhibition that opened with fanfare in the Louvre in September 1819. It awarded medals of gold, silver, and bronze. Among the products displayed were Baccarat crystal and porcelain from Sèvres and, among the mechanical exhibits, a Wolf steam engine. Christian compared an exposition of machines to a natural history museum; it displayed the "fruits of human industry" but should also illustrate conceptions of economy and principles of good manufacture.[18]

The collections over which Christian presided gave him another venue through which to communicate his vision of the organization of industrial society. The mechanisms provided a visual illustration of mechanics' intelligence and were a powerful demonstration of what could be done with motive force; they transmitted it, changed its direction, regulated or suspended it, accelerated or retarded it, and diffused it or stored it up for a single decisive action.[19] Rochefoucauld, as chair of the conservatory's governing board, had, upon Christian's appointment, insisted that he begin to prepare a systematic catalog of the collections. The first volume appeared in 1817. The catalog remained an accomplishment of which Christian was particularly proud.[20] He placed it in the tradition of the *Encyclopédie* of Jean d'Alembert and noted that the collections revealed the essence of modern society by displaying how tastes and needs intersected with diverse methods of production. But it was not enough merely to illustrate all the different ways of producing things; the catalog aimed also to identify the means that were the most direct and most economical, or the most "perfected."[21]

Christian's associates Rochefoucauld and Carnot shared a belief in the press as a powerful tool of education—they had seen it at work in the popular publications of the Revolution.[22] Christian himself turned to publication as a way to educate both cultivated men and working mechanics in the industrial sciences. In addition to the conservatory's catalog, he wrote a treatise on preparing linen and hemp without retting, a cumbersome process of soaking the stalks to loosen the fibers; oversaw publication of numerous volumes describing and illustrating patented processes, mechanisms, and machines; published a critical, theoretical study of industrial operations and a multivolume survey of industrial mechanics; and produced *L'Industriel,* a serial devoted to developments in industrial and mechanical operations.[23]

L'Industriel illustrates Christian's approach: to discuss real machines in actual use. He inaugurated the serial in 1826, and it ran until 1830, using a format established in its first year. Each issue opened with a discussion of a general mechanical principle: the definition of a motor in the first number, later on the concepts of force, mass, or weight or the qualities of atmospheric air that might affect the performance of a steam engine. A description of one or more machines already in use followed, identified by their location and the names of their builders or owners and illustrated in engraved plates bound at the back. Each number concluded with a list of recently granted patents in both France and England. Christian opened the premier issue of *L'Industriel* with a statement of its goal: to help the luminaries of science penetrate the establishments of industry.[24]

Claude-Pierre Molard, Christian's predecessor as conservatory director, had called technical drawing "the sixth sense of the worker," and Christian had in his institution one of the most celebrated draftsmen and engravers of his age, namely Leblanc, instructor and, after 1829, professor at the conservatory's Petite école. As Leblanc described it, the point of drawing was not only to give a visual portrait of a machine but to create "an exact understanding of the composition of all types of machines."[25] Christian wrote of mechanisms, like words, being arranged in a mechanical combination, just as words were arranged in a literary composition, and he described Leblanc's engravings as "another language" for communicating mechanical information.[26] Leblanc's engravings graced several of Christian's publications and were a powerful technique for the effective transmission of information about machines. Well-executed drawings made it possible to understand a machine and its parts without handling it and aided in controlling their construction. They also helped publicize machines of proven merit, selected for their solid construction, good working histories, and economy of production.[27]

Transforming Labor

To Christian the most effective machines were dynamic agents of social transformation. They economized on human labor, freeing it for other work and replacing strenuous, machinelike, and repetitive tasks with ones requiring intelligence and judicious movement. He proposed a theory of industrial operations founded on an analysis of industrial labor and centered on the concept of economy, qualified by consideration of a machine's social utility. Christian wrote of perfected machines and processes, in which not only their quantified performance mattered but also their social effects.

France was in the midst of turbulent industrialization, and the introduction of machinery into a wide variety of manufactures spurred resistance. France experienced machine-breaking riots in the 1780s and 1790s, especially against new textile machinery, as workers decried new machines as intrusive and feared the destruction of their livelihoods and their communities. Many blamed the zeal for mechanization on the years of war, which had kept industry shorthanded. Roger Magraw described a number of incidents of protest: outcries from shoemakers in 1817; protests by shawl makers in Picardy in 1818, who succeeded in having machinery removed; the worries of a furniture maker in Faubourg St-Antoine who feared for the safety of his mechanized saws. Resistance was greatest in the woolen industry; a shearing machine, the "Great Cropper," was greeted with a riot, dumped into the river, and attacked by master shearers with their scissors when it arrived at Vienne (Isère) in 1819. It was erected anyway. Violence in France was never as great as in England, where troops put down an uprising in the north in 1811–12 and the House of Lords passed the death penalty for machine breakers, but it continued sporadically for nearly fifty years.[28]

French workers did find themselves in precarious positions, often newly at the mercy of the market and consumers' changing desires. A formal and protective system of guilds and apprenticeships had been a casualty of the Revolution, but workers had not embraced economic individualism in its stead. In fact, they cherished their corporations, found ways to maintain them secretly, and resisted the doctrines of laissez-faire.[29] Workers remained within a subsistence economy throughout the Restoration, facing seasonal unemployment and fluctuations in work occasioned by the fickle tastes of those in the market for French luxury goods. Jacques Étienne Bédé, an accomplished chair turner who worked in Paris until he was jailed in 1821 for his part in a dispute with workshop owners, described how the vagaries of taste influenced his work: turners hauled in the wood they used themselves, and when tastes changed to favor styles using unfinished wood, the task became harder and more dangerous and required carrying unfinished planks down twisting stairs and sometimes bringing in "enormous hunks of wood, often consisting of entire tree trunks and enormous beams."[30] Even the proprietors of workshops were not spared uncertainties.[31] As William Sewell put it, French industrialization produced "an abundance of discontented artisans," who later became the mainspring of the working-class movement.[32]

Christian published his industrial philosophy in 1819 in *Vues sur le système général des operations industrielles, ou plan de Technonomie,* a tract examining both exuberance and distress about machines and how they might be deployed in emerging industrial society. *Plan de Technonomie* appeared the same year that the Conservatoire des Arts et Métiers adopted its new system of public lectures and near the end of a period of anti-machine riots. Although it has long been overshadowed by the conservatory's well-known lecture program, historians have begun to give Christian's critique an important place in the history of industrial philosophy.[33] The tract was unusual in Christian's oeuvre. With the exception of a foldout table on the last page, it was all text, and it lacked the elegant and precise plates by Leblanc that would later grace *Traité de mécanique industrielle* and the serial *L'Industriel. Plan de Technonomie* was a single volume, slim and small.

The work was a critical and theoretical engagement with French industrial concerns of the early nineteenth century. Central was a concept of industrial operations, characterized by the idea of linkages between mechanisms in a productive chain. It was one of the first theories of industrial work in a tradition usually dated to Andrew Ure and Charles Babbage some two decades later. The theory was an integration of technology and economy, conceived as a successor to the concept of technology developed in the descriptive project of Johannes Beckmann. *Plan de Technonomie* was not an inventory of machines or a survey of materials and sites of production but an explicit and complex introduction to a science of industrial production, including both a new periodization of the history of technology and a comprehensive and unprecedented system of classification of machines. It distinguished between crafts or trades, in which productive effect was achieved directly by hand or by a hand equipped with a tool, and manufactures, in which production was accomplished by a combination of mechanical operations exercised in concert. Christian also addressed issues of consumption, the circulation of capital, and the moral dimensions of a factory system. He acknowledged machinery as a powerful stimulant to prosperity.[34]

Plan de Technonomie offered a view of economy not through ceaseless division of labor but through a careful distinction between work that demanded human dexterity and intelligence and work that required repetitive movements, even if highly skilled. Christian believed his theory to be of importance to industry, especially as encapsulated in an analytical table he described to Rochefoucauld in 1817 and which appears at the end of the text and makes his distinction between crafts and manufactures clear. Christian believed the distinction offered a new approach to the problem of resistance to machines by identifying where machinery could be introduced without degrading the types or conditions of work.[35] He recognized two forms of degradation: exhausting work that left a laborer fatigued and repetitive, rote work that dulled the intellect. Machines could supply force, and they could replace repetitive, determined motions. More intricate work would continue to require human hands and human attention.

An analysis of motion, both of machines and of the human hand, was thus central to Christian's philosophy. He described motive force as a sort of substance that could be stored, conserved, or purchased, often at a dear price.[36] It was real, he argued, neither abstract nor ideal. Motors generated primitive movement, thus providing essential motive force; mechanisms then guided the motion to appropriate points of application, bringing force to bear on the original materials from which needs were met and desires fulfilled: fibers of cotton or wool to be stretched and twisted into thread, grains such as wheat to be ground into meal or flour. But different types of fibers could not be spun in the same way, even wool needed to be spun differently for a pile either short or deep, and cereals could not be ground like corn and other oily grains. Every mechanical operation used motive force but to impart varying motions and often to entirely different effect.[37]

The hand moved in especially labyrinthine and complex ways, and Christian doubted that a mechanism could ever mimic it with either exactitude or economy. Hands represented more generally a worker's bodily involvement in the productive process. Physical labor was not limited to handicrafts, and tasks such as heavy lifting or pushing required moving the legs. But much skilled work did originate in movements of the hands, arms, and powerful muscles of the upper torso, as in blacksmithing or carpentry, while the hips and legs formed a stable, stationary undergirding for the strenuous work done above.

The quality of judgment, in addition to intricacies of movement, was one of the reasons human handiwork would never be entirely replaced by machines. Handiwork remained especially important in French industrialization. France

had lost its colonies and their ready markets during the Revolutionary and Na-
poleonic Wars, and French entrepreneurs increasingly concentrated on high-
quality and well-crafted goods, commanding higher prices at export. Skilled craft
workers were valued, and Christian saw in their good taste a particular asset. The
high opinion of French taste was not new; the century before the Parliament of
Paris had described with pride the renown of French merchandise, "its taste, its
beauty, its finesse, its solidity, the correctness of its design, the perfection of its
execution, the quality of its raw materials."[38] Artisans were the mainstay of
French industry. Even late in the nineteenth century small manufactures employed twice as many people as did larger enterprises.[39]

Christian found important differences between repetitive, determined motions and varying and indeterminate ones. Machines were useful only for tasks
that required either repetitive movements or movements whose variations could
be specified exactly and in advance. Also important was the distinction between
work demanding sustained attention and tasks that could be left to a well-
constructed machine. Mechanical science remained powerless to perform work
demanding human judgment and continued oversight. Such work was to be further distinguished from tasks requiring skill or habit. In tailoring work the movements of the hand might change at each instant, constantly guided by the intelligence; polishing marble, in contrast, required continuous and repetitive
movements. Polishing was skilled work, for pressure had to be distributed evenly,
but once it had been learned the task became rote.[40]

Movements involved in spinning illustrate the interplay of mechanical movements and human handiwork and attention that were so important to Christian;
it was an example he cited often. Spinning required careful control of motion,
demanding a strong enough torque in the right direction to impart a twist to the
fibers but not so much or at such an angle as to snap the yarn. Mechanisms did
not have a light enough touch to replace individual spinning wheels until well
into the eighteenth century, and even then they still required careful attention
from their operators, who had to engage the roving of fiber with the mechanism
that would draw it out and twist it, wind the thread onto a spindle or bobbin, and
all the while regulate the amount of attenuation and twist required for varying
degrees of fineness. It required judgment and wide-ranging movements to reach
spindles or rovings deep within the machinery or to reconnect broken threads.[41]
Spinning by hand with a wheel made no intellectual demands; though it was
skilled work, it depended on habitude rather than judicious movement. Christian
believed that mechanical spinning was different: it required sustained attention

to a combination of mechanical devices and knowledge of how they worked. It stretched the intellect and trained the mind.[42]

The great diversity of mechanisms available was a material manifestation of the intelligence of the mechanics who had made them, and Christian believed that the same intelligence could be developed in their operators. Machines would not replace handiwork that required varied motions and sustained attention, and where machines were introduced, they would require new forms of intelligence and dexterity. This had been the case historically, Christian argued. When people turned to nonhuman sources of power for rote, repetitive tasks such as pushing, pulling, pounding, or turning handles and cranks, it freed their hands for other work. Animal power as well as wind, water, and steam offered new possibilities for human labor.[43] Industry would create new forms of work, demanding laborers of increased intelligence for preparing materials and operating machinery. Christian cited the work of preparing rags for paper making and mining for needed metals and coal, and he alluded to an infinity of tasks not yet known. His examples do not strike us as liberating, underground mining in particular having been recognized as especially debilitating, but they did offer scope for movement and discretion, and for Christian these were their important features.[44]

An outgrowth of Christian's emphasis on mentally demanding work was a criticism of Adam Smith's theory of the division of labor. Even when it spared workers fatigue and difficulty, subdividing tasks might undermine their capacity for intelligent reasoning. Christian did agree with Smith that people were naturally motivated to improve their conditions; it supported his argument that workers be given tasks that engaged them fully. But he believed that the division of labor had reached its limit when it sought further economies by making tasks more repetitive or rote.[45] Machines could be used abusively, and the social transformation that Christian foresaw required paternalistic direction by heads of enterprise and oversight by the state. The chief of industry would treat his enterprise like a family, exercising manufacturing economies like he would exercise economies at home in the domestic space. This demanded concern not only for expenses but also for the well-being of family members and workers, through "a paternal solicitude" and continual action on their behalf, in an atmosphere of "an inflexible good faith." The government would concern itself with matters of political economy, necessary but beyond Christian's scope: issues of importation, taxes and duties, and the circulation of capital. Above all the importance of human labor must be recognized and the ways that machines could enhance it.[46]

Christian's concern for preserving mentally challenging and physically varied human handiwork illustrates the role of effective machines in his analysis of emerging industrial society: effective machines were identified not only by their productive performance but also by their social impact. In a period of persistent and sometimes violent resistance to new methods of industry Christian offered a theory of how machines could transform labor to the worker's benefit. Not only would they replace forms of labor requiring brute human strength; they would also create new and intellectually demanding forms to replace rote and repetitive toil. This was a dynamic notion of effect, concerned not with quantity of work but with its quality.

Conserving Motive Force

Christian's general concept of the effect of industry was dynamic and transformative; his specific concept of the effect of machines was static. He used a prototype of efficiency in describing the perfection or advantages of machines, in a ratio balancing their mechanical effects against the fuel or combustible they used. But the static nature of his prototype lay not only in a balanced equation but also in Christian's very definition of a machine. Machines, in contrast to workers, were not allowed a wide range of movements. Only machines whose movements were rigorously controlled would display economy in performance, and only they could be called perfected. Christian's advice was practical and straightforward: in order to build the most effective machine, hold everything still, except the necessary moving parts.

This approach allowed Christian to circumvent the continuing dispute over how motion itself should be characterized and quantified, a question still unresolved since Smeaton's time. Like Smeaton, Christian saw important differences between the sciences and the mechanical arts, and he rejected the description of industrial mechanics as an applied science. His object was real working machines, not machines in the abstract, and he distinguished his field from the analytical tradition of rational mechanics represented by mathematicians such as Cauchy, his old rival at the École Polytechnique. His was a practical analysis, aimed at helping mechanics ensure that machines performed as intended. Christian's emphasis on the control of motion was thus disciplinary, intended to guarantee a desired outcome.

Perfection described the principal advantage of the best machines: they produced not only what was desired but did so with the greatest economy. Christian had introduced the issue in *Plan de Technonomie*, but it also figured prominently in his textbook on industrial mechanics, *Traité de mécanique industrielle,* published in four volumes between 1822 and 1825. He used as an example the symbol of industry, the steam engine. All other things being equal, the most advantageous engine produced "the greatest mechanical effect, while using the least amount of fuel (combustible)."[47] A similar concept of the perfected machine appeared in Leblanc's work, in the folio volumes of his *Recueil des machines, instrumens et appareils qui servent à l'économie rurale et industrielle* (1826), in which he presented engravings of the best working machines he could find. He described a high-pressure steam engine of 25 horsepower, which possessed "great advantages" through both the "simplicity of its movement" and its economical use of fuel. A hydraulic turbine constructed by Fourneyron displayed "remarkable advances" due to the facility of its workings, which increased useful effect and improved conservation.[48] A nascent efficiency percentage figured in Leblanc's description of another turbine, by Chartres mechanic Fontaine: it possessed "incontestable advantages," allowing it to return a useful effect of "70 to 75 per 100" with a chute of 1.4 meters.[49]

This prototype of efficiency also appeared in patent applications, underscoring the widespread use of the ratio between mechanical effect and resources used as an index of machine economy. Of the forty-three steam engines patented in France between 1792 and 1824, 37 percent were for improvements described as perfections, producing greater economy in performance, either in combustible used or effect produced, or economy of materials in construction.[50] In 1822 John Leberecht-Steinhauser sought a patent for a steam engine with parallel movement but without a balancing beam; the principal point of perfection in his design was its "transmission of the force of the steam" to the working parts without any intermediary pieces, which reduced friction and resulted in a "grande économie."[51] Englishman Edward Hall sought patent for a high-pressure steam engine whose perfection lay in its economizing of combustible; he claimed a greater economy by half than achieved by engines of Boulton & Watt.[52] Paris mechanic Jean-Christian Dietz sought a patent of perfection for a machine called the steam wheel, whose economy of motion and materials made it "very important for the French government and the factories of Royaume."[53]

These measurements of machine performance did not confront the thorny conceptual difficulties of assessing motion, in motion. They assessed static states before and after motion had occurred, much as did the Franklin Institute in its waterwheel trials of the 1830s, using the widely accepted measurement of work as the product of mass times distance moved. The issues Smeaton had found distracting were still in play,[54] and terms that have become basic to mechanical science were still in flux: Christian himself did not distinguish between *force* and *power,* for example. The static concept of work allowed Christian to offer counsel on practical actions despite theoretical confusion.

In articulating the distinction between the practical tradition of industrial mechanics and the mathematical tradition of rational mechanics, Christian described a conception of movement that compared successive static states. Industrial mechanics used devices to accumulate or disperse motive force as a way of producing effects. Measuring the immediate and momentary action of a motor was not the point; mechanical operations required collecting that action, saving it or transmitting it, in part or in full, in different directions and over different distances, and transforming it into work.[55] Measuring forces in equilibrium or at an instant was of no use in industry, especially if the calculations assumed ideal conditions or a theoretical balance between acting and reacting forces. It mattered little "to know that pounds could be made equal to a column of water, or to the piston of a steam engine, or what burden a man or a horse could carry or support without having to change place."[56] This was not movement, Christian argued, for the idea of equilibrium supposed nothing happened. In industrial mechanics things had to change places. Christian particularly resisted abstract measurements of time, such as an "instant," finding them of no practical value. Knowing "an exact expression of [a motor's] mechanical force" at an instant was only momentary knowledge.[57] Mechanics needed knowledge of successive movements, how motion was transmitted, and the effect that would result.

Christian's emphasis on working machines falls within the tradition of engineering economics that Ted Porter has described as allied "more closely with accounting than with high theory."[58] A traditional method of accounting for the performance of manufactories had been the ratio of items produced to the number of workers employed, but new devices made it less meaningful. Jean-Antoine Chaptal proposed in 1819 that the ratio of products to machines would be a better measure of the performance of an enterprise.[59] Christian himself described following motion through a chain of operations as a method of accounting. "Motive force is a reality," he wrote; "it is a sort of raw material, that one can, so to speak store up, that one should economize, that one always buys and for which one often pays dearly."[60]

It was held "rigorously immobile" so that the moving force was directed solely to the work at hand: this was Christian's definition of a true machine. Machines were subject to extraneous movements such as vibrations and shaking, which were "forms of motion like any other" but foreign to the effect a mechanic wished to produce. The first step in accounting for them was to fix the machine firmly in place, to bolt it to the floor, attach it to a beam, or build into it sufficiently firm support.[61] Motion was also lost through friction, more difficult to combat because it was necessary for the transmission of force from one part of a machine to another. A mechanic seeking to improve a reasonably good machine should look at how its pieces related to each other to determine "the relative degree of perfection" that was possible.[62] Improvement would be seen when all extraneous motion was countered by an appropriate combination of mechanisms; the fewest parts arranged in the most economical fashion conserved the most motive force. Four characteristics of motion were important, whether or not the motion was desired: its direction; its path from point of generation to point of application; its type, whether rotating or linear and whether in a horizontal or vertical plane; and its regularity. Christian provided ways to deal with all of them.[63]

A perfected machine thus conserved the moving force given it by an engine or motor through the proper arrangement of supporting parts to minimize loss of motion through friction, vibration, or shaking. Success in improving a machine depended upon the ability to foresee how its various parts would interact so that its effect could be predicted with some accuracy; success required not only conserving motion but controlling it. Machines, in contrast to workers, were not to be allowed a wide range of movements. Only machines whose motion was rigorously controlled would display economy in performance; only they could be called perfected.

This control of motion was disciplinary, designed to guarantee that machines would do what they were intended to do. Christian did not extend this discipline to workers: he neither described ways to control how they moved nor analyzed how the use of machines may have helped do so. There is ample evidence that heads of workshops and enterprises did install machines in an effort to discipline those who worked for them, even when the devices produced inferior products,[64] but such cases did not enter into Christian's analysis.

A visual demonstration of the basic principles of motion control appears in Leblanc's 1828 study of the English spinning mill outside Compiègne. Leblanc's philosophy employed drawing as a method of analyzing machines, mentally taking them down and then building them up again. Drawings allowed mechanics who were interested in improving a machine's effectiveness to study it in a static state, to see how its pieces would connect and what path its motion would take. Leblanc's engravings of the mill displayed the intricacies of its many pieces, illustrated separately and together, the final plate showing the full system in place.

The penultimate plate illustrates the system of motion control, the bolts and bearings that fix the apparatus in place. Molard, under-director of the Conservatoire des Arts et Métiers, wrote the book's foreword, in which he described the sys-

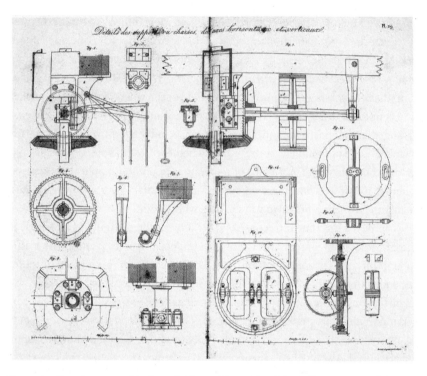

Elements of motion control in the spinning mill outside of Compiègne, engraved by Leblanc. V. Leblanc, *Nouveau système complet de filature de coton, usité en Angleterre et importé en France par la compagnie établie à Ourscamp, prés Compiègne* (Paris: Bachelier, 1828), pl. 29. Courtesy of the Harlan Hatcher Graduate Library, University of Michigan.

tem of supports as an important English advance. It allowed limited and precise motions while rendering the bulk of the machine immobile, lessening the costs of friction and the loss of motive power. The juxtaposition of moving and stationary parts was at the root of modern spinning methods, seen in devices that carded wool by pulling one moving card across another one that was held immobile, or in the spinning jenny, in which carded fibers were held under static tension while being spun. Leblanc's engraving illustrates the tension between the need for motion and the need to control it: heavy bolts secure cast-iron seatings to a stout beam, allowing motion only through a series of elegant and tiny bearings. The heavy workings of the mill were held in a stationary embrace.[65]

Christian's concept of the perfected machine was an intellectual construction erected at the crossroads of mechanical practice and mechanics as theory. It was part of the emerging field of industrial mechanics, and it employed a broad notion of effect that encompassed not only quantitative measures of machine performance but also wider and qualitative considerations of the role of machines in society. It illustrates the complexity of the intellectual traditions from which the formal concept of efficiency would emerge.

Christian's work, like Smeaton's and the Franklin Institute's, employed a prototype of efficiency. But unlike them, Christian embedded discussions of machine performance within a thoroughgoing philosophy of industrial society. His concept of effectiveness combined motion control in machines with intelligent and educated motion by their operators and by other workers who were engaged in handicrafts or small manufactures. Christian's notion of the effect of machines was two-pronged. Effective machines would release human laborers from repetitive and strenuous toil and would free them to do work that stretched their intellects and their bodies. They would transform human drudgery. But the most effective machines, the most perfect, would themselves be perfect drudges, working mechanically and by rote, their parts following repetitive and predetermined paths. They were to remain primarily static or immobile.

Human work and machine work were thus fundamentally different categories to Christian, and he assessed their effects in different ways. He quantified machine motion, using a prototype of efficiency that balanced a machine's mechanical effect against the fuel or motive force it had used. He did not quantify how workers moved, and he was explicit that workers should not be required to move in machinelike ways. Human movement was something different; it would not be calculated nor controlled or predetermined. But human work could be transformed. Christian, writing from the platform of the Conservatoire des Arts et Métiers at a time when machines were widely regarded with suspicion, argued that machines themselves should be the agents of that transformation.

CHAPTER THREE

Economy of Nature

Darwin, Marshall, and the Costs of Efficiency

Efficiency came to have two meanings in the mid-nineteenth century: a conservative meaning, allied with ideas of balance, and a progressive one, corresponding to the dynamics of progress.[1] These meanings grew out of concerns about the limits to natural resources, and their waste and loss, and also as a result of important shifts in how natural, physical, and economic processes were described, away from concepts of equilibrium, or balance, and toward dynamic, or growth-oriented, understandings. Efficiency in its static and conservative sense reflected concerns about resources and their limits, by measuring how to operate within their constraints. Efficiency in its dynamic sense also reflected concerns about natural limits and waste and became a tool for overcoming those limits through organization and management. Naturalist Charles Darwin and economist Alfred Marshall, two prominent Victorian thinkers, used these meanings of *efficiency* in their most well-known works: they wrote of efficiency in the balance of nature and in the economy. Just as Darwin found a correlation between the size of a genus and its success at adaptation in the efficiency of natural selection, Marshall found a correlation between firm size and innovation in the efficiency of business management. This chapter looks at how efficiency is treated in the famous first edition of Darwin's *On the Origin of Species by Means of Natural Selection, or the Preservation of Favoured Races in the Struggle for Life* (1859) and the eighth and final edition of Marshall's seminal *Principles of Economics*, the first edition of which was published in 1890. It illustrates the widening use of the concept and its integration into diverse intellectual constructions beyond the mechanical and practical.[2]

Both Darwin and Marshall conceived of efficiency as a function of diversity; both saw diversification essential at all levels of the hierarchies they studied. Darwin saw it everywhere, from the level of structures and functions within the individual organism to the branching of groups, which he called "divergence." Marshall saw the increasing specialization and subdivision of labor as essential to the success of large business enterprises. But the diversification of which Darwin wrote differed significantly from Marshall's. Business efficiency, to Marshall, resulted from the deliberate integration by business managers of diverse and specialized labor. Darwin described natural selection as efficient, but it did not deliberately integrate the variety of individual differences that, when accumulated, gave rise to divergent forms of life. Natural selection acted efficiently, but it did not deliberately organize. The increasing variety in uses of efficiency thus came to include both its intentional use and its chance uncovering.

Efficiency: Theoretical and Mechanical

The background to the widening use of efficiency was its formalization as a concept in engineering and physics. The young French theorist Sadi Carnot constructed a prototype of theoretical efficiency, writing of motive force, or power, in a publication of 1828. Sadi Carnot and his father, Lazare Carnot, were part of a French engineering tradition situated between purely rational theory and the practical design and construction of working machines. Lazare Carnot had derived conservation principles from work on the physics of impact and applied to the construction of turbines and waterwheels,[3] and Sadi Carnot extended this analysis to the steam engine, basing his concept of ideal efficiency on the relationship between moving power and heat. Rather than consuming quantities of the imponderable fluid caloric, steam engines derived their power from the movement of caloric between a hot body and a cooler one, or from heat source to heat sink. Motive power derived solely from the temperature differential, with maximum efficiency being dependent upon maximum differential. His was a mechanical law, the younger Carnot argued, to which all sources of power were subject: "is it possible to conceive the phenomenon of heat and electricity as due to anything else than some kind of motion of the body, and as such should they not be subjected to the general laws of mechanics?"[4] At issue in nineteenth-century thermodynamics was precisely this question, whether caloric should be subject to the laws of mechanics and, if so, whether and in what form it was conserved.[5]

Sadi Carnot's work received scant attention during his lifetime and had little bearing on practical mechanical considerations. The model for theoretical efficiency was a machine, the steam engine, but Carnot, in the rational mechanical tradition, had analyzed it in the ideal case. He had not measured heat and power in actual machines. Later in the century William John Macquorn Rankine, Regious Professor of Civil Engineering and Mechanics at Glasgow from 1855 to 1872, developed a conception of efficiency linking it to thermal and mechanical science and employing it in larger arguments about the economy. Rankine used efficiency to create a respectable academic presence for engineering science. He refined the concept and used it to link precision in engineering analysis with economic gain, creating "pure science *regulated by economy.*"[6] Rankine distinguished between natural and commercial economies; science investigated phenomena "connected with the general economy of nature," whereas engineering was linked to economy in the financial sense.[7] Both economies incorporated the idea of conservation, in commercial terms through the need to economize or to compensate for waste by increased productivity. The definition of *efficiency* offered by Rankine thus extended the concept of balance and stasis but with an eye to management rather than analysis.

In Rankine's definition balance in the economy of nature made use of the emerging concept of conservation of energy. James Prescott Joule's theory of the mechanical equivalent of heat postulated that motion and heat were interconvertible and was later used to link Carnot's discussion of the theoretical efficiency of a temperature differential to the thermal and mechanical efficiency of a moving machine. The simultaneous discovery of the conservation of energy in the 1850s indicates widespread interest in the relationship between heat, motion, and an interchangeable, or equivalent, force.[8] Rankine and his colleagues defined *efficiency* as a machine's "economy of power or energy; that is the ratio of the useful work [it] performed to the energy exerted."[9] Energy was conserved in the economy of nature, but useful work performed was defined in a broader and expansive context, that of economic progress and growth. In Glasgow, an industrializing city, useful work interested both the university's governing board and Rankine's colleagues on the faculty and at the Glasgow Philosophical Society.[10] Rankine sought a precise mathematical definition of efficiency to reflect both the scientific credentials of engineering science and the ability of a trained engineer to serve the needs of an economy increasingly concerned with growth and waste. Rankine used the term *efficiency,* linking the physical sciences with a widespread interest in fostering industry, to consolidate his claim for the academic respectability and economic necessity of engineering science.[11]

Mechanical and thermal efficiencies were more than measures of the internal economy of a machine; they were also tools for managing and introducing machines into the economy. Armed with practical knowledge, engineers could conveniently and profitably manage and direct "the expenditure of work." The service of economy "directed and controlled by efficiency" was Rankine's motif. As a ratio comparing means needed "to produce an effect, and the means actually employed," Rankine's definition was quantitative; it had the added appeal of the notion of perfection.[12] A perfectly efficient ratio of resources used to work performed would be one, or unity, 100 percent of what had been expended having been conserved. The ratio of efficiency thus measured the degree of perfection of its object; the lower the efficiency of a machine or process, the less perfect and more wasteful it was. A scientifically trained engineer could make improvements aimed at "diminish[ing] the lost work so as to make the efficiency [of a machine] approximate to unity." Efficiency thus served as a measure of engineering progress. It created a hierarchy of perfection and an accurate method of measurement that pointed out to the manager where improvements might be made.[13]

Efficiency thus had a technical and limited meaning, which measured the balance of effort or energy put into a machine against effect produced, and a general and broad one, concerned with larger economic performance measured not in terms of inputs and outputs but in how effectively it promoted the public good. In the seminal works of Charles Darwin and Alfred Marshall efficiency mediated between mundane elements, which could be treated mechanically, and larger, less well-defined natural or economic systems.

From Static Balance to Progressive Change

During the nineteenth century the economy of nature, which included living beings and particulate matter, came to be described as a dynamic, progressive system, rather than a static, balanced one. The idea of economy in nature was rarely discussed by naturalists before the nineteenth century but became commonplace within a few decades of Georges Cuvier's reconstruction of the fossilized skeletal remains of animals he believed had become extinct. Prevailing descriptions had emphasized the balance of nature, although the term *economy* was used, as in Carolus Linnaeus's work, to designate an organizational scheme. Extinction disturbed the stability of the balance of nature. Although Scotsman William Smelling had in 1791 argued that "no species is ever exhausted—the balance between gain and loss is perpetually preserved,"[14] within a generation Robert Chambers was writing of the bloody extermination of species for their own good. At midcentury Charles Darwin, in his own evolutionary theory, made the shift away from the static conception of nature in balance complete.[15]

Physicists, engineers, and political economists had also begun to argue for a progressive conception to replace the idea of a static balance. A new concept of dynamism and change characterized work in the physical sciences in Victorian Britain and replaced the former view of natural or economic systems as tending toward equilibrium and resulting in a static or balanced state. Relations between scientific men and industrialists, as Norton Wise and Crosbie Smith put it, "provide[d] the ground of real interests where 'economy of nature' and 'political economy' often merged in meaning."[16]

As extinction had upset the balance for naturalists, machines such as the steam engine upset the balance for physical scientists and economists. Division of labor, which for economists since Adam Smith had been a primary means of producing wealth, could not account for the increased productivity made possible by machines. Dividing tasks did not allow for fundamental and progressive change. The English economist David Ricardo saw a form of "self-generating" progress in technology that could lead to a stationary state via the law of decreasing returns but could not be adequately treated in the static conception of a balanced economy.[17]

Darwin followed an earlier generation of naturalists, particularly the uniformitarian geologist Charles Lyell in arguing that the world of nature was not static but, rather, was subject to geological change, climatic change, change of other sorts, of a gradual nature; Marshall argued that the human economy was not static but experienced continual change in rents, wages, and organization.[18] Both men not only attempted to understand the mechanisms of such change but also discerned a positive direction in that change, Darwin implicitly, Marshall explicitly. For Darwin change increased diversity among living organisms; for Marshall change made it possible seriously to entertain the possibility of eradicating poverty.[19]

Darwin and Marshall used efficiency to distinguish objects susceptible to management, such as people, processes, and machines, from the agents of management. Efficiency describing a balanced, controlled, or stable state indicated an object to be managed; in contrast, efficiency describing the general public or natural good indicated an agent of management.

Charles Darwin: Efficiency and Growth

In the first edition of the *Origin of Species* Darwin used a principle and its consequence, natural selection and divergence of character, to explain how from simple beginnings and through a natural economy that appeared balanced and uniform, nature could generate "endless forms most beautiful and most wonderful."[20] He portrayed natural selection as an efficient principle of management with divergence as its consequence. In Darwin's metaphor nature managed varieties by selecting those that would survive. If they survived, more specialized varieties (the most divergent) would become distinct species. Natural selection acted; variation, which provided the increasing variety of forms among which to select, gave nature something to act upon.

The Dynamic Efficiency of Natural Selection

In the *Origin's* first chapter Darwin drew an analogy between something familiar, agricultural breeding, and his own challenging theory. Both selection and divergence occurred under domestication: the stimulation of variability by the exposure of organic beings to new conditions of life and the process of selection by which people adapted plants and animals for their own purposes. People did not actually produce variations, Darwin acknowledged, providing instead conditions in which they might appear and taking advantage of their appearance to select and accumulate those they found desirable. Darwin described human selection as a powerful agent in creating useful domestic breeds, some with the "character of natural species." The same selective principles acted efficiently in nature. People selecting among available variations had efficiently produced breeds that functioned as natural species.[21]

Darwin also compared the efficiencies of human and natural selection. The efficiency of natural selection buttressed Darwin's argument for the analogy of natural to human selection. "There is no obvious reason," he wrote, "why the principles which have acted so efficiently under domestication should not have acted under nature." Of all the causes of change he surveyed, he found natural selection "by far the predominant power," acting "more efficiently" than methodical and rapid purpose-driven choice. Human selection was feeble in comparison.[22] Natural selection was efficient in a dynamic sense, more far-reaching than the rearrangement of parts and functions that characterized changes within the balance model.

Among the factors that most persuaded Darwin of the efficiency of natural selection was the case of neuter insects: "But I am bound to confess, that, with all my faith in this principle, I should never have anticipated that natural selection could have been efficient in so high a degree, had not the case of these neuter insects convinced me of the fact." He called the problem of neuter insects "by far the most serious special difficulty" his theory had encountered.[23] In describing natural selection as efficient with regard to neuter insects, Darwin used a broad conception of efficiency linked to management of a large and changing system, moving beyond the metaphor of balance.

Sterile or sexless insects posed two particular difficulties for the theory of natural selection: Darwin had to explain how it had rendered only some insects within a breeding population sterile as well as how it could account for differences in structure and instinct between the sterile insects and their parents. Various species of British and foreign ants that employed castes of workers, soldiers, or guardians provided Darwin an opportunity to study this phenomenon. A working ant both differed greatly from its parents and was sterile, thus without means of transmitting its differences to succeeding generations.[24]

Darwin approached this problem by expanding the scope of natural selection. "Selection may be applied to the family," he wrote, in addition to individuals; what was advantageous to the community, in the case of social insects, might well be preserved in the struggle for existence. The greatest difficulty was accounting for the development of structurally and behaviorally distinct castes within insect populations, in which even neuters differed significantly from each other, sometimes "being as distinct from each other, as are any two species of the same genus, or rather as any two genera of the same family." Darwin proposed that nature selected parents that produced useful neuters and saw evidence of this process in various species of British ants, in which neuters exhibited a wide gradation of characteristics, of which only a few fell into the intermediate range. The West African driver ant, *Anomma,* also exhibited early development of neuter castes, its castes graduating "insensibly into each other."[25] Darwin argued that nature would select families producing the most extreme varieties of neuters and, over time, would eliminate families producing more moderate forms; social insects exhibiting distinct castes of neuters were thus more highly developed.

Natural selection, in the case of neuter insects, was dynamic and managerial; it selected the families with the most distinct neuter castes and launched them on a trajectory of further adaptation and advance. There is an important point here. Divergence resulted from the action of natural selection, from the accumulation of minute physical differences. Divergence did not act, however; natural selection did. The diversification that gave rise to divergence was efficient in the stable sense of showing a balance and was the object of natural selection, the managing agency; it described the relative similarity or dissimilarity of living beings, which, as natural selection acted, simply continued to diverge, moving farther apart but still competing along the same balance.

Diversification in neuter insects, like division of labor, was something to be put to use, to be managed; neuters were the only way to achieve what Darwin termed the "perfect" division of labor. Had the neuters been fertile, they would have bred and, according to the theories of inheritance available to Darwin, diluted through blending the distinct features and instincts that so differentiated them from their parents. Distinct and neuter working castes were analogous to the division of labor that marked advanced civilization, Darwin argued: "We can see how useful their production may have been to a social community of insects, on the same principle that the division of labour is useful to civilised man."[26] The division of labor was historically associated with economic models emphasizing balance rather than growth. Malthusian theory, holding that population grows exponentially while resources grow linearly, was itself a model of balance: population growth was checked by limited resources, or, in other words, available resources limited and thus balanced population. Adam Smith's liberal economic model, with which Darwin was familiar and which emphasized competition as the method of maximizing human happiness, employed the concept of balance in arguments for the division of labor, tasks given up by one worker being taken on by another.[27]

Static Efficiency of Divergence

Darwin developed the concept of divergence to explain how the extreme variety of individual living beings gave rise to distinct forms of life, natural selection being unable to explain how species became more than "variations on a single theme." Without divergence Darwin was unable to explain how species became genera.[28] Innumerable balancing acts, in which extreme variation in an individual or family tipped the scale in its favor, combined, through the agency of natural selection, to form the branching tree of life. By diversifying in character from their immediate competitors, individuals fit into places in the economy of nature still incompletely exploited. There were gaps in nature; it was not a continuum. Darwin wrote explicitly of balance: "Under nature, the slightest difference of structure or constitution may well turn the nicely balanced scale in the struggle for life, and so be preserved."[29] Consider the following passage, in which Darwin described the relationship between diversification and natural selection: "More individuals are born than can possibly survive. A grain in the balance will determine which individual shall live and which shall die,—which variety or species shall increase in number, and which shall decrease, or finally become extinct . . . The slightest advantage in one being, at any age or during any season, over those with which it comes into competition . . . will turn the balance."[30] The grain in the balance could be the grain of pollen that "the efficiency of insect-agency" carried to a hybrid rhododendron or a rudimentary organ "perfectly efficient" for something else.[31]

Silvan Schweber argues that Darwin's concept of divergence was analogous to concepts of the division of labor, especially as employed in the work of Adam Smith and the French naturalist Henri Milne-Edwards. In this sense, particularly as it operated on individuals, diversity could be considered an extension of the static economy of nature.[32] Janet Browne argues that Darwin employed the notion of division of labor to explain how a geographic area could support the greatest amount of life. Darwin's biomass argument, that an area supporting more varieties also supports more life in the aggregate, turned on his example of an experiment in which plots of ground sown with seeds of one species were compared with plots sown with the seeds of a variety of species and genera. The area seeded with more variety yielded "a greater number of plants and a greater weight of dry herbage."[33] Darwin argued that "the greatest amount of life can be supported by great diversification of structure," noting that farmers rotated crops on this principle, whereas nature practiced "simultaneous rotation."[34] He employed the economic concept of division of labor and explicitly cited Milne-Edwards on this point. The full concept of divergence, however, depended not only on variety, or division of labor, but on variety in its most extreme forms and its accumulation into distinct forms of life.[35]

Natural Selection as an Agent of Progress

Natural selection had broad scope as it made use of diversity of character. Darwin argued that selection scrutinized "the whole constitution, structure, and habits of each creature—favouring the good and rejecting the bad," working so that the characteristics of all living beings tended "to progress towards perfection." He made the management metaphor clear: natural selection undertook the "work of improvement" of species by selecting "only for [the good] of the being which she tends."[36]

Darwin employed the language of optimization, associated with the limits of balance, in a progressive framework of cause and effect with a definite agent of change. He would have rejected the argument that descent with modification had a purpose or goal, but he did not reject causal argument.[37] Natural selection was a causal agent of change, as was nature itself: "natural selection is daily and hourly scrutinising, throughout the world, every variation, even the slightest; rejecting that which is bad, preserving and adding up all that is good." "The face of Nature," Darwin wrote, "may be compared to a yielding surface, with ten thousand sharp wedges packed close together and driven inwards by incessant blows." Natural selection drove the hammer that struck the wedge.[38] Natural selection was efficient and dynamic because it was an agent of cause operating in the largest possible theater, that of all living things.

A distinction is important here. Darwin's language allowed interpretations that were inconsistent with chance.[39] He wrote of natural selection as though it were a conscious entity, scrutinizing and selecting, and this caused him great difficulties.[40] The efficiency of the principle did not depend, however, on Darwin's own suggestion of purpose or design. It depended instead on issues of agency and progress. Natural selection was an efficient agent, even without purpose.

To illustrate his theory of divergence Darwin included in the *Origin* a now famous diagram, "Diagram of Divergence of Taxa." The full concept of divergence relied upon the balance being turned in favor of individuals possessing divergent but advantageous characteristics, and the accumulation of those differences to the point that they marked a group as more or less distinct. To account for the rise of genera, their characteristics had to diverge continually. At first glance Darwin's diagram appears to depict progress, or certainly change rather than stasis. But the diagram illustrated more than the simple concept of divergence; it also depicted the action of natural selection in preserving and accumulating the benefits derived from that divergence. Darwin's diagram is a picture of the aggregate effect of innumerable individual instances of diversity and selection, over tens of thousands of generations, and it illustrates both the static efficiency of divergence and the dynamic efficiency of natural selection.

This idea can perhaps best be illustrated with a mechanical exercise. Consider for a moment Darwin's discussion of balances and wedges. A lever, or wedge, can be analyzed in the static terms of opposing but equal forces. The amount of force pushing down on one end of the lever is equal to the force generated at the other end. The equation is balanced. In Darwin's concept of divergence, when the balance was tipped in favor of an individual living being, it was also tipped against a competitor, and thus the metaphor of balance remained. But at some point the infinitely small incidents of balancing added up to something more, and the system became dynamic. Although Darwin offered extinction as a corollary to balance speciation, his diagram moved beyond the static notions of the economy of nature supposed by a balancing of speciation and extinction. The extreme varieties of which Darwin wrote, the "accidents" and "sports," did not average themselves out around a mean. They were selected and moved up on his diagram, while the average shifted. Natural selection drove wedges down onto the balance of nature, eliminating some species and advancing favored others. Extinction

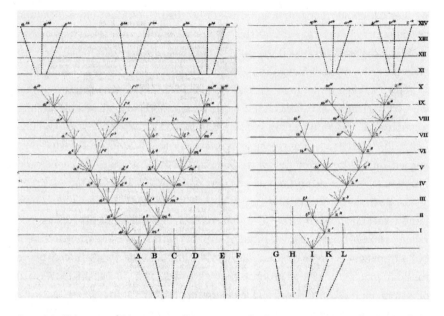

Darwin's "Diagram of Divergence of Taxa." Natural selection, acting on individuals of extreme variation, produced the branching forms of life depicted in the diagram. The outer dotted lines represent the most divergent variations, the shorter, fan-like dotted lines represent varieties that were not successful in reproducing. Dotted lines reaching any of the horizontal lines were sufficiently well marked as varieties that Darwin considered them incipient species. Heavier dotted lines at the top of the diagram represent distinct species descended from species A and species I (and species F, which descended without modification), including species that may form genera or subgenera. The two groups of species descended from I were more closely related, having diverged more recently, than the three groups of species descended from A. The principle of divergence was efficient, for example, in producing at least eight distinct species from the single species A. Charles Darwin, *On the Origin of Species by Means of Natural Selection, or the Preservation of Favoured Races in the Struggle for Life,* facsimile of 1st ed. (1859; rpt., Cambridge: Harvard University Press, 1964), endpapers.

and extermination, offered as balancing principles, did not prevent the increasing perfection of adaptation of which Darwin wrote. Darwin did not illustrate a dynamic of extinction broad enough to balance descent with modification. Exterminated individuals and species simply disappeared, while selected individuals and species were integrated into an evolving system, only perhaps to disappear in turn.

Nature, driving in a wedge, upset the balance and launched selected individuals and species on a potentially infinite trajectory of further adaptation. But nature's wedge, as Darwin depicted it, did not balance that trajectory. It did not strike with sufficient force. Individuals and species selected against, those that became extinct, simply disappeared from the diagram; they were not driven on an opposite but equal path. Darwin's tree of life had no roots, no mirror image. It did not, in the end, illustrate balance.

Nor did Darwin give extinction or extermination a balancing rhetorical weight. John Fleming had devoted an entire volume of his work on zoology to factors limiting life,[41] and Robert Chambers wrote forcefully of the evils and misery required "to bring the rest of animated beings to a proper balance."[42] Darwin wrote often of extinction and extermination but rarely of misery or evil. Donald Worster noted Darwin's unwillingness to fully embrace the implications of extinction.[43] Species often exterminated their parents in the *Origin*, for example, but dispassionately. Descendants represented on the upper levels of his diagram, Darwin wrote, "will have taken the places of, and thus exterminated, not only their parents (A) and (I), but likewise some of the original species which were most nearly related to their parents." In discussing classification, he wrote that "there is a constant tendency in the forms which are increasing in number and diverging in character, to supplant and exterminate the less divergent, the less improved, and preceding forms." This comparison enabled him to distinguish growing genera from those that were dwindling, the older genera composed of more widely divergent forms but no longer increasing in numbers. Next, however, Darwin turned his attention to growth and life, further explaining the advance of the branches in his diagram. Although he spoke of balance in nature, what Darwin was really describing was progress. Although natural selection would not provide a living being with something injurious to itself, it might favor a characteristic injurious to a being's competitors. Even within species themselves, Darwin argued, "if a fair balance be struck between the good and evil caused by each part, each will be found on the whole advantageous."[44] Darwin chose not to emphasize extinction and death but, instead, the "grandeur" of his view of life, its beauty and wonder.

The final point is not that Darwin's theory of descent with modification was progressive or dynamic. A popular conception is that it was, although historians and biologists have refuted that view. Stephen J. Gould argued that Darwin refused to use comparative terms such as *higher* or *lower* to describe different species. Dov Ospovat, in contrast, considers Darwin's theory progressive, as were other developmental theories of his time.[45] Darwin described a dynamic and managed system. His mechanical metaphors, his continued invocation of the balance in particular, and his characterization of the principal components of his theory, natural selection and divergence, as efficient combined to produce not a stable and self-regulating system but one of dynamism and movement. Natural selection was an active agent, daily and hourly scrutinizing, driving in wedges, efficient not in balancing forces but in upsetting them. Darwin did speak of efficiency in balance, in which the face of nature remained largely uniform despite innumerable and minute struggles, but he generally characterized it as progressive and dynamic and characterized it visually as a tree. "Divergence of Taxa" was a diagram of growth.

Alfred Marshall: Efficiency as a Function of Organization

Insects were important for Darwin—the efficient bee minimizing its waste of wax and the castes of servant ants who were neuter and yet continually replaced. Ants and bees also played a role in Alfred Marshall's classic *Principles of Economics,* in which he used them to illustrate the relationship between division of labor and economic efficiency. The fourth book of Marshall's *Principles,* discussed here in his final, eighth edition, was devoted to the role of organization in economic growth. In it Marshall, too, employed *efficiency* in two senses: to describe the balanced and stable functioning of the division of labor and to describe the larger benefits accruing to those skilled in managing that labor.

Marshall has been credited with introducing Darwinian concepts into the increasingly professionalized field of political economy; Joseph Schumpeter called Marshall "one of the first economists to realize that economics is an evolutionary science."[46] Others have questioned how greatly Marshall was influenced by Darwin, arguing that Marshall rarely cited Darwin explicitly and that many ideas now considered Darwinian had, in the late nineteenth century, a wide currency beyond evolutionary debates.[47] Marshall did, however, draw analogies between economics and the physical and biological sciences and suggested that economics supported a combination of the methods of both.[48]

Marshall, a founder of neoclassical economics, trained in mathematics at Cambridge and returned there as professor of political economy in 1885. By the time of his death, in 1924, most chairs of political economy in Britain were occupied by his students; through his teaching and the *Principles* Marshall dominated British economics for more than forty years.[49] The *Principles* became a classic almost immediately upon its publication in 1890; Marshall's publisher, Macmillan, used it to lead a campaign against falling resale prices in the book trade. George Stigler considered *Principles* the second greatest work in the history of economics, after Smith's *Wealth of Nations.*[50] *Principles* replaced John Stuart Mill's *Principles of Political Economy* as the favored economics textbook in Britain.[51]

Marshall distinguished between statics and dynamics in economics, arguing that the subject of economics was dynamic but that the method of analysis should be static. "The central idea of economics, even when its Foundations alone are under discussion must be that of living force and movement," Marshall wrote. "The main concern of economics is thus with human beings who are impelled, for good and evil, to change and progress." But social, economic, and natural forces were so numerous that only a few could be considered at a time. Primary relationships should be isolated, he argued, other forces reduced to inaction "by the phrase 'all other things being equal.'" Suggestions of economic rest, which included the "hypothesis of a Stationary state," were provisional and only described a way in which "sensible men have dealt from time immemorial with every difficult problem of ordinary life." Static isolation of a specific economic relationship allowed it to be closely analyzed, although the economist should not lose sight of the changing world to which it was linked. Treating variables *provisionally* as constant did not mean treating them as inert; it meant instead temporarily ignoring them, Marshall argued.[52]

Efficiency and Organization

Both dynamic and stable efficiency figured in the *Principles*. For Marshall efficiency was a function of the increasing subdivision of labor, as managed by those with high degrees of business skill. It described the fittest and most highly developed organizations and economies; their growth depended, in turn, on the stable and efficient functioning of a well-managed base of labor and machinery. The goal of efficiency was worth pursuing, Marshall argued, even though it demanded sacrifice on the part of its members. Economics, for Marshall, as for other neoclassical economists, was a matter of rational choice and planning.[53]

Efficiency, Marshall reasoned, was both the result of organization and its further stimulus, noting that "writers on social science from the time of Plato downwards have delighted to dwell on the increased efficiency which labor derives from organization."[54] In a provisionally static state, Marshall argued, the law of increasing return, which depended on increasing efficiency, would be balanced by the law of diminishing return, in which nature resisted the increased production of raw materials. According to Marshall, the law of increasing return "may be worded thus:—An increase of labour and capital leads generally to improved organization, which increases the efficiency of the work of labour and capital."[55] A firm's or industry's increasing returns could override diminishing returns as it became increasingly larger; the larger the business organization, the more it depended on its own internal economy. That internal economy was subject to management and efficiency through organization in a way that natural resources were not. The large firm was the most highly organized and thus the most efficient and was administered by the ablest of men; this was efficiency in the sense of progress or growth. Such progress could be observed in the increasing organization that marked recent economic history. Adam Smith had argued that the organization and division of labor had allowed increasing numbers of people to be supported on limited acreage; it had also served to preserve the most highly organized races against the pressure of population, which "weed[ed] out" those that showed a "want of organization."[56]

The division of labor stimulated efficiency by differentiating functions and increasing the specialization of various tasks. The division of labor observed in biology and economics was an expression of the same fundamental principle. But there was more to efficiency than differentiation of tasks, for organization also included their reintegration by management. Marshall argued that organization, as a function of capital, was a distinct agent of production in its own right, the central question relating industrial organization to efficiency being how and whether business management itself should be subdivided. Business skill differed from industrial skill—there was no question that the efficiency of manual labor could be increased, he noted, but whether the mental labors of management could be divided was a more complex question.[57] It is at this point that Marshall's discussion of efficiency took on two forms, distinguishing between the efficiency of the managed and the efficiency of the manager.

Marshall thus developed a hierarchy of labor that distinguished tasks that could be managed in the interests of efficiency from those that contributed to overall economic efficiency but could not be managed precisely. He separated tasks that could be identified, prescribed, and divided, and which thus lent themselves to management, from tasks of a different order, which required mental agility and skill and were less easily differentiated and managed.

"There is no doubt," Marshall wrote, "that greatly increased efficiency can be attained through division of labour in those occupations in which there is much demand for mere manual skill." Division of labor, a characteristic of the economic application of balance, was certain to stimulate efficiency in cases in which physical labor alone was in question. Degrees of skill separated the backward from the progressive races, Marshall believed, just as it separated the higher and lower grades of labor. In machine weaving, for example, the higher grades were able to weave more complicated patterns, while "most of those who work in the lower grades have not 'the stuff in them' that is required for weaving with several colours." Workers acquired skill, primarily through residence among a people in whom habits of responsibility and carefulness had developed over several generations and only secondarily through training. Indeed, skill was often not recognized among a skilled people, who had become so familiar with the handling of materials or machinery that they considered such tasks routine and mechanical. Marshall reckoned that perhaps one-tenth of the world population had the "mental and moral faculties" to perform labor classed as unskilled by civilized peoples; they would require at least two generations of steady training to be fitted for it. Even among manufacturing populations Marshall believed that only a small portion were able to perform tasks considered monotonous and thus requiring little skill.[58] Efficiency was a function of the environment inhabited by the supply of labor.

The character of the laboring classes was, to Marshall, more important than training or practice. Qualities of mental strength and self-mastery allowed a "general command over the use of one's fingers," an important element of industrial efficiency; training could promote general manual skill and thus had value, so long as it was not too specialized. Manual skill that could not be transferred to different tasks or put to use in a different occupation was losing importance in the developing industrial economy, in which a great industrial people was marked by the ability to "act promptly and show resource" and to "accommodate oneself quickly to changes."[59] Marshall distinguished this general ability from what he called "specialized ability," which was characterized by manual dexterity and familiarity with the particular processes and materials required in the individual trades.

These passages in the *Principles* reflected Marshall's years of thinking on efficiency, race, and innovation. The progressive development of complex organization characterized the history of the natural world; that same progressive development also characterized the history of civilization and the economic world. Increased specialization and increased efficiency, through technological or mechanical ingenuity and the extension of human skill, characterized the most advanced nations. They also characterized the most adaptable and strongest firms within a given economy.[60] It was through competition that the greatest nations had achieved success, and through adaptation to their geographic surroundings and the economic possibilities such surroundings might support. In lecturing on the future place of the United States in the course of world history, Marshall described the view he had taken from Hegel: that civilizations historically developed from the East to the West, from the Far East to the Mediterranean to western Europe and now (1875–76) across the Atlantic. What most marked a developing economic power was mechanical inventiveness and general ingenuity, both illustrating adaptability to the natural economic environment. The United States had this in spades, according to Marshall, who foresaw continued British decline, at least in terms of economic competition.[61] The point here is simple but crucial: technological inventiveness was a necessary component of adaptation and thus of economic survival. This was for Marshall a historical argument, and he characterized such adaptations as efficient.

Mechanical inventiveness and mechanical skill were not innate traits, however; they were traits of a population, and individuals learned them socially. In a series of notes on efficiency and populations taken in the decade preceding the *Principles,* Marshall suggested that highly skilled peoples became so used to their skill that they ceased to recognize it and thus mistakenly considered much regular labor as unskilled. Over a couple of generations the members of a population or society could be taught inventiveness and skill, just as they could be taught the industrial virtues: industriousness, punctuality, forbearance, and loyalty. General education could also stimulate the broader aptitudes that brought about true creativity and efficiency.[62]

Efficiency of Labor and Population

Marshall kept notes on efficiency under two headings: labor and population. The efficiency of labor (rates of population increase, age of marriage and mother's fertility, and indications of a superior breed of population) and that of population (characteristics of workmen in different nations, the effects of technical education on work performance, and the character of British workmen) had at their core the same theme: the relative industrial aptitude of the various European races. Marshall's interest in what he called the "efficiency of population" led him to determine not only the shared characteristics of the most effective industrial workers but also which nationalities produced them. Heredity was not an adequate explanation; as Marshall saw it, heredity could not account for physical, mental, and moral characteristics and their combinations. He kept notes on instances of improvement through diet and nurture, looking for other explanations for national differences. In an 1883 report by the Anthropometric Committee of the British Association comparing average heights and weights of the Scottish, Welsh, English, and Irish, he found "a difference of 5 inches . . . between the best and worst nurtured classes of children of corresponding ages."[63]

Marshall's wife, Mary Paley Marshall, contributed notes on the eugenicist Francis G. Galton and efficiency, especially his comments on family size and cultural, political, and economic success. Larger families had higher moral standards, Galton argued, and Mary Marshall copied down his explanation (owing to "internal discipline, family pride and criticism") alongside Galton's citation of Sir William Gull, who had remarked that the strongest candidates for the India Civil Service came from large families. Galton's method of determining flourishing and vigorous families was similar to that used by Darwin to distinguish growing from waning genera. Large size, numerous living relations, and several generations of occupational success by its men indicated a vital and growing family. Small size, fewer relations, instances of occupational distress, and early deaths indicated a weakening family.

Mary Marshall's notes detailed ways in which the Royal Commission on Technical Instruction's 1884 report confirmed the superiority of English over Continental workers, despite the more systematic education available across the Channel. She recorded several examples: Continental dyeing firms were often led by scientifically trained chemists, but in the application of chemical processes English industry had met with equal commercial success and was as innovative "in new and important discoveries." Continental soda works were managed by better-trained men, though English industry did not suffer because "with one or two exceptions everything in the way of important improvements in the Leblanc alkali process has originated in England." Compulsory military service taught German workers "order[,] regularity[,] obedience[,] and power of working effectively with others . . . [but] Germans don't get through so much work as English," having less "go," or "staying power." This the notes attributed to a combination of race, climate, and the competitive spirit of the English, although German workmen were considered here to be more docile, trustworthy, and intelligent, and Berlin, which had once been "a drunken city," had through education taught its workers to adopt a more temperate life. The commission quoted a Swiss, who confirmed that "the English are at the lead of all the workmen he has ever seen," including those of Italy, France, Germany, and his native Switzerland. The English had greater practical knowledge and more mechanical genius than their Continental counterparts, even without technical education. Various descriptions peppered the notes: the Englishman had greater physical endurance, more "all around capacity," superior physical strength, and superior natural aptitude. A reporter from Zurich found these superiorities to be "mainly due to [the] fact that [the Englishman] lives better," although, he observed, English factory girls made poor wives, and Swiss women were better at household management. The English manager of an engineering works in Belgium estimated that English workers could do 20 percent more work in the same time as Frenchmen or Germans and were more energetic and of greater strength, due to the bracing English climate (which was tolerable the year-round) and to the greater concentration of mechanical industries. Also, English workmen were "brought up on more animal food." Germans had taken the lead in the chemical industries, including production of artificial dyes and ammonium processes, and Belgium had perfected deep-mine ventilation. Mulhouse, France, produced the best printed fabrics, France's soft woolen textiles were unmatched, and Continental artisans were better trained in drawing and various primary subjects. Nevertheless, England remained "head of [the] industrial world." Evidence of this came in the form of Bessemer's steel-refining process, which, along with Gilchrist's innovations in the uses of the commonest iron ore, had cut the price of steel by 90 percent during the previous decade. The list of evidence continued: Perkins aniline dyes, Lister's silk process, land and submarine telegraphy, new applications of water pressure, a revolution in the alkali trades, and growing Continental reliance on English machines and English technical journals.[64]

Efficiency and Society

The Marshalls' notes on efficiency display the insistence that efficiency began at the level of a population or society. Although efficiency could be recognized in individuals, its reward, or its highest use, could be found in the greater collective: the society or race.

Marshall had the opportunity to observe efficiency in different societies at first hand during his travels in the United States. In the summer of 1875 he undertook a voyage to New York with a legacy of 250 pounds left to him by his uncle Charles. This year marked the end of what his biographer Peter Groenewegen has called Marshall's "economic apprenticeship" and was one of the first tasks he set himself after having settled the matter of his vocation. He wished to observe a dynamic and growing economy in its infant stages. Mary Paley Marshall later recalled that "he never spent money so well. It was not so much what he learnt there as that he got to know what things he wanted to learn. He was able to see the coming supremacy of the U.S., and to know its causes and the direction it would take." His American trip also showed to him the coming difficulties England would face through declining competitiveness.[65]

While traveling, Marshall prepared a series of notes for use in his later lectures. He toured the docks of New York and visited a stove works and a reaping and mowing machine works in Albany; the Bessemer works between Albany and Troy and, nearby, a horseshoe works and a huge cotton mill; the Waltham, Mas-

sachusetts, watch factory (the largest in the world), the Cambridge, Massachu-
setts, Mason & Hamblins organ factory, and the Chickering piano works; a nail
works and glass maker near Pittsburgh; and the mines of Virginia City. He toured
the New Lebanon settlement of the Shakers and the Oneida community, stayed
with Charles William Eliot, president of Harvard, supped with Ralph Waldo Emerson, and walked and talked with William Graham Sumner. He spent his entire
legacy and returned early in the fall. Marshall was most impressed by what he
called the American sense of "Go," something the Canadians unfortunately did
not share. He saw it everywhere he went: in an ingenious adaptation of old
machines in the horseshoe works (which he unfortunately did not describe); in
clever swindles at the silver mines of Virginia City; and in the frequent report that
Americans would not stay at a job very long because they were always casting
about for something better.

Marshall also commented on two things that would be important to his conception of efficiency: the race or nationality of workers and managers and who
was responsible for mechanical innovations. Observing race or nationality, Marshall commented that the Americans themselves seemed to be more generally
skilled and present in almost all types of trades. Members of other nationalities—
Germans, Irish, French—congregated in specific trades. Irish workers made the
moldings at the stove works, for example, and French workers designed patterns
for the cotton mill. Germans in particular were not as efficient as the Irish, being
"(i) more given to drink and (ii) 'not so apt to ask reason why.'"[66] Here was an
early example of the hierarchy of skills that Marshall would develop in the *Principles*. The more general and adaptable the skill, the closer to management.

Marshall also commented on management in American concerns, though
obliquely, through his observations on mechanical innovation. Innovation came
from the managers, not from the laboring men on the floor. If workers did invent
something, they did not know how to patent it or how to estimate its worth. Here,
in a microcosm, was Marshall's argument about general ability and economic
efficiency—illustrated in a conversation Marshall had with the young man showing him around the Albany stove works. Marshall commented that the man was
"obviously in authority." Marshall asked, "Are many inventions made by the
men?" "No," the man answered, "at least as a rule: they are almost entirely made
in the shop. One & only one invention of importance has been made here by a
working man. He got $100 for it; He might have sold it, if he had waited & man-
aged properly, for $20,000." Marshall and his guide dismissed this single in-
stance of mechanical inventiveness as an example of a worker's mismanage-
ment. Marshall's guide continued: "If inventions are made by a working man at

all, it is in general not by those brought up in the trade who have stuck to it, but by men who have had a good education to start with, & who have knocked about in the world & shifted from one occupation to another."[67] Mechanical inventions, Marshall noted in this conversation, did not come from those highly skilled, trained, and specialized. Inventiveness belonged to the more generally skilled.

Marshall investigated this question again when touring the Bessemer works between Albany and Troy. He interrogated his host, who was quite proud of the mechanical ingenuity displayed in the plant, on how such inventiveness had come about. Marshall asserted, "Then the large number of your working classes who have the habit of thinking for themselves, does indirectly promote inventions, to a greater extent than it appears." He then qualified his statement: "If many inventions are not made by those who work with their hands, many are made by those who have worked with their hands." And with this his host agreed.[68] Thus, we have another example of Marshall's early conceptions of a hierarchy of labor and skill and the primary importance of management. These are the elements that fifteen years later formed the core of his notions of efficiency: sacrifice and management.

An Efficiency of Progress

Marshall distinguished between two types of efficiency: first, efficiency that could be managed, that of manual labor and tasks that could be analyzed in a stationary state; and second, the efficiency of management, which was measured in the size of a firm but was otherwise difficult to assess. A basic level of efficiency could be achieved through the differentiation of function, but for true growth these functions had to be reintegrated into the industrial organism. Integration was the task of business managers, who did highly skilled and mental work and whose efficiency was more difficult to assess than that of the manual worker. This second type could be called an "efficiency of progress," characterized not by breaking even but by growth. That growth could only be achieved by the sacrifice of autonomy by all but the most highly skilled at management.[69]

General ability was the higher form of ability, in Marshall's view, because it allowed the adaptation to changed conditions that was necessary for survival in a competitive economy. Technical education still primarily concerned itself with knowledge of the particular trades and thus produced minds that were unable to adapt to changing business conditions. Although many tasks could be performed equally well by educated and uneducated workers, general education stimulated the production of material wealth by generating more trustworthy, steady, and intelligent laborers. In addition, general education was more likely to develop the artistic and creative faculties that were becoming "a chief factor of industrial efficiency."[70]

Management of business was the highest level on Marshall's hierarchy of labor. A "specialized body of employers [or] business men" intervened between manual workers and consumers by directing production so that it supplied human wants in the most effective way. The highest grade of business ability was the most varied in application; the able businessman, Marshall argued, was able to adapt in an increasingly complex business environment. He displayed this ability during periods of misfortune as well as of success, and, according to Marshall, proof of his ability was found in the size of the business he owned: "There is a far more close correspondence between the ability of business men and the size of the businesses which they own than at first sight would appear probable."[71]

Industrial efficiency, Marshall reasoned, depended upon the health and strength of the population and upon whether it was able to withstand the pressures of its own growth. He cited Smith's argument that the more highly organized a society, the more people it could comfortably support; the most productive division of labor occurred in a system of individual sacrifice within a relatively flexible social system. The rigid organization of labor that characterized slaveholding and oppressive societies was a major cause of inefficiency, as was the lack of food, which occasioned a general weakening of the population. True industrial efficiency, Marshall believed, was collective; it could not be achieved without a healthy and stable supply of labor, which in turn depended upon a society able to accommodate its needs.[72]

There was a cost to be borne in the pursuit of efficiency, however, and here we return to the ants and the bees. Marshall noted that with increasing differentiation of function, "each part gets to be less and less self-sufficient, to depend for its wellbeing more and more on other parts." Of particular importance was the integrating or managing function, to which individuals were to respond with sacrifice and a noble patriotism. "A tribal sense of duty" could be seen in the impulses of the social insects such as the bees and ants; the most noble races had a highly developed ideal of "deliberate and therefore moral, self-sacrifice . . . in which the individual is most willing to sacrifice himself for the benefit of those around him, and which are consequently the best adapted collectively to make use of their environment." Marshall called this "the standard of family and race duty," in which individuals were willing to perform various tasks for society without direct recompense to themselves. It operated in the wild, where even predatory animals would exert themselves without direct gain on behalf of their offspring. Societies whose members were willing to sacrifice for the good of the collective held an advantage in war against other societies whose members did not, and in struggles against famine and disease. "Thus the struggle for existence causes in the long run those races of men to survive in which the individual is most willing to sacrifice himself for the benefit of those around him," Marshall wrote, "and which are consequently the best adapted collectively to make use of their environment."[73]

Shortly after returning from the United States, Marshall gave a public lecture in Cambridge entitled "Some Features of American Industry." He took as his theme the relationship between human character and economic conditions, arguing that the ethical development of individuals and societies could not be divorced from the economic situation that underlay them. In particular, the ethical progress of any particular society depended on the daily activities and business of life.[74] Marshall was to return to this theme, the relationship of ethical development and daily economic business, throughout his career. It would thus be unfair to characterize him as a social Darwinist who considered economic extinction desirable in and of itself. Marshall believed that people could be educated, and he thought nurture was more important than heredity. He had originally turned to political economy while looking for ways to overcome poverty.[75]

For Marshall, however, ethical values were tied to economies. One central feature that distinguished different economies, as it also distinguished the different societies that went along with them, was efficiency. Higher rates of efficiency marked the most adaptive and dynamic economies and societies. Efficiency, in turn, depended on an extreme division of labor, individual sacrifice, and, finally and most crucially, competent management. People and societies that could neither achieve nor maintain efficiency would be subject, in an increasingly sophisticated international marketplace, to management by people and societies that could. For Marshall this meant deep worries about Britain's loss of competitiveness and the subsequent threat to its economic autonomy.

Two forms of efficiency, one describing stable and balanced conditions, the other a system dynamic and growing, operated in Marshall's work. The distinction can also be seen in Darwin's *Origin,* in which natural selection acted efficiently in both senses. The distinction is not minor. Darwin's use of the concept of efficiency highlights questions about natural selection as an active agent and whether the view of life he described was progressive; both questions agitated his contemporaries and occupy historians and evolutionary thinkers today. Marshall's use of efficiency highlights a distinction between management and labor and a connection between organization and monopoly, relationships of economic and social control that have interested activists and social thinkers for more than the past century. The question of efficiency also brings into relief a central tension in both works. In Darwin the tension lay between bitter individual struggle at the level of the balance and the overall progression of life. In Marshall the tension lay between increasing specialization and social progress and the threat of monopoly. For both men efficiency meant increasing and sophisticated organization necessarily accompanied by sacrifice: the death and extinction of less-adapted and less-specialized organic beings and the loss of autonomy by those engaged in all but the most mentally demanding forms of labor.

There was, however, a crucial distinction between them. For Marshall efficiency grew out of rational choice and the deliberate integration, by a manager, of various skills, materials, and resources into an organization or firm. For Darwin natural selection acted, but there was no agent of deliberate integration to organize the results. Both the *Origin of Species* and the *Principles of Economics* also invoked dynamic processes that marked progress in life and raised the possibility of eradicating poverty. They illustrate the increasing range of the concept of efficiency and its affinity not only with mechanical matters but also with the broadest of intellectual concerns.

CHAPTER FOUR

Balance and Transformation

Technical and Popular Efficiency
in the Progressive Era United States

Clara Corrigan was the most efficient young woman in the folding room of the bleachery in the midwinter of 1909. A year later that was not so clear. Frances Grogan, Theresa Liepert, and Mary Livsey had caught up with her. They folded lengths of cloth longwise and in book style, or like an accordion. They might do fifteen different kinds of things in a day, perhaps some cutting or knotting or else making smaller and more intricate folds. Watching them was Henry Gantt, efficiency consultant and mechanical engineer, disciple of Frederick Winslow Taylor (with whom he later broke), graduate of Johns Hopkins University and Stevens Institute of Technology. The women probably worked in the Sayles Bleacheries of Saylesville, Rhode Island, where Gantt had done some consulting work. Gantt created a graph of the women's efficiency: black meant efficiency, red (here shown as gray) meant failure. The graph is an early example of a Gantt chart, a tool of engineering management still in wide use. It illustrates a combination of technical and personal efficiencies and dates from the Progressive era, when concern for efficiency was at its height—efficiency in industry, in government and public life, and in personal affairs.[1]

Engineering Magazine published Gantt's chart in 1910; its province was technical efficiency. The *Independent*, a well-respected journal of reform, was also interested in efficiency, though of a personal kind. Technical and personal efficiency in *Engineering Magazine* and the *Independent*, despite their apparent differences, shared a number of similarities, evident in Gantt's chart and including a reliance on control and a goal of uniformity and stability. They were both efficiencies of balance, uniting complex and shifting components into a reliable, regular, and stable platform. Efficiency was, in both, a matter of management: in *Engineering Magazine*, of machines, systems, or people; in the *Independent*, of oneself. But they differed in an important respect. Discussion of efficiency in *Engineering Magazine* was sober and matter-of-fact; in the *Independent* it evoked a dynamic transformation, of self and society.

Efficiency unfolded into a variety of uses as the concept became more than an intellectual and analytical construction and was elaborated as a useful, workable tool. The liberal sweep of the examples here—from engine tests embodying efficiency's mechanical core to new and arbitrary measures of work performance to injunctions to personal discipline—makes manifest the malleability, changeability, and reach of the concept. At the same time they demonstrate its contrariness, in the contrast between its conservative operation as a gauge of engine performance and its invocation as a tool of radical personal reconstruction.

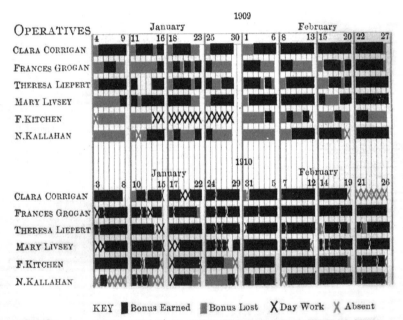

KEY ■ Bonus Earned ▤ Bonus Lost ✗ Day Work ⋎ Absent

CHART VII.   BONUS RECORD OF GIRLS WORKING IN A FOLDING ROOM.
The upper half shows eight weeks in 1909. The lower half, the corresponding 8 weeks in 1910. Remembering that red means bonus lost and black bonus earned, the improvement in 12 months becomes strikingly apparent.

Gantt's chart of young women working in the folding room of a bleachery, probably the Sayles Bleacheries of Saylesville, RI. Black indicates 100 percent efficiency for the day, gray (red in original) indicates failure. Henry L. Gantt, "Fixing Habits of Industry," *Engineering Magazine* 39 (1910): 172. Courtesy of the Minnesota Library Access Center, University of Minnesota; photo by Nancy Johnson.

Progressive Era Efficiency

Gantt wrote in *Engineering Magazine* in 1910 that true efficiency was to be found "in a state of equilibrium which is stable." He meant a balanced efficiency, measured less in increased productivity than in guaranteed regularity and reliability. "Any prosperity not based on efficiency is on an unstable foundation," he wrote. Edward Earle Purinton, efficiency consultant and popular writer, wrote in the *Independent* in 1914, in response to a query about efficiency unrewarded on the job, that one's "first duty is to be the finest possible worker in your present situation." He, too, meant a balanced efficiency: predictable, reliable, dependable, and well regulated.[2]

Gantt and Purinton were efficiency experts in an era of efficiency experts, in a time when people preached efficiency like a gospel. Concern for efficiency, and a variety of techniques for measuring and applying it, marked many of the reform efforts that characterized American and European history before the Great Depression, reforms spurred by worries over the effects of industrialization and urbanization in a changing international order and expressed in efficiency movements tied to national health, governmental reform, military prowess, and protection of empire, nation, or race. In the United States the Progressive era was a time of intellectual, social, and political turbulence, often dated to the presidencies of Theodore Roosevelt and Woodrow Wilson, occasioned by significant changes in how people lived, and marked by changes in how they ordered themselves and their society. Businesses and industries were becoming bigger (railroads were the prime example), and they employed increasing numbers of people, including children, who worked long hours for little pay and without protection in cases of accident. Economic depression and strikes in the 1890s had convinced many people that government should take a greater role in regulating economic and business life, and increasing evidence of corruption convinced them that government needed also to better order itself. Individualistic and local forms of government and business gave way to an increasingly bureaucratic order, managed more by experts whose authority came from scientific and technical knowledge and less by leaders distinguished through personal qualities such as integrity and diligence.[3]

Efficiency became a tool of reform, as one of a cluster of "languages of discontent" that underlay Progressive thought and practice.[4] Historians have described a widespread enthusiasm for efficiency in the United States in this period, especially between 1910, when hearings on rates and waste in railroad operations brought efficiency and scientific management to wide public attention, and 1914, with the start of the European war. Efficiency became ubiquitous. It described not only technical matters, such as the thermal economy of an engine, but personal ones as well: finding a mate, rearing children, or respecting one's boss. It also spoke of honest government, fair industrial policy, and far-sighted resource management. Technical features such as quantification and calculation jumbled together with social, governmental, and personal concerns to produce a word resonating technical expertise, personal integrity, and good government. Efficiency expressed both sober qualities of hard and patient work and enormous hopes for remaking society and the world.[5]

Enthusiasm for efficiency truly exploded after the Eastern Rates case of 1910, in which the Interstate Commerce Commission held hearings on proposed railroad tariff increases. Attorney Louis Brandeis, representing businesses, argued against the rate increase and summoned efficiency consultant Harrington Emerson as an expert. Emerson was an important link between the expertise of scientific management and the larger efficiency movement.[6] He did not hold an engineering degree but was a longtime advisor to the Santa Fe Railroad; Brandeis had asked Taylor to recommend someone to testify, Taylor had recommended Gantt, who had some experience with the Canadian Pacific Railroad, and Gantt in turn had recommended Emerson. Emerson became Brandeis's star witness. Emerson had recently written that through their inefficiency the railroads wasted $300 million a year; at the hearing Brandeis subtly adjusted the figure, arguing that the railroads had no need of increased revenues, for through efficiency they could save "one million dollars a day."[7] It was a pleasing phrase and a catchy headline, and it fired the public's imagination. Efficiency soon described diets, appliances, child rearing and education, clothing design, exercise, saving money, curing neurasthenia, and the path to global dominance. Efficiency itself became a prominent feature of advertising—of motors, soaps, and breakfast foods—and appeared in daily newspapers, home economics journals, and mechanics' magazines. By 1915 the word *efficiency* was plastered everywhere—in headlines, advertisements, editorials, business manuals, and church bulletins.

Efficiency had both grand and prosaic roots, associated with management and organization but also with enterprises as spectacular as the heavenly order itself. *Engineering Magazine* and the *Independent* represented two traditions of efficiency—one technical, allied with the science of energy and machine design and testing practices, the other general, associated with economy and prudent management. The traditions were not entirely independent. Engineers and physicists created technical efficiency in the nineteenth century by giving mathematical definition to a term taken from discussions of organization and management; W. J. M. Rankine, for example, had created definitions of mechanical and thermal efficiency at the University of Glasgow in the 1850s (see chap. 3). Ratios of output to input had been used earlier, most famously in the mid-eighteenth century by the English engineer John Smeaton in an analysis of the mechanical power of waterwheels (see chap. 1); Smeaton had not used the term *efficiency* and had indeed expressed some question about what term he should use. *Efficiency* entered the engineering and physics lexicon in the mid-nineteenth century, alongside efforts to quantify the performance of machines. The term's association with economy and oversight preceded its use in physics and engineering, extending back to early modern discussions of the wisdom and efficiency of the creator, who had established and maintained the order of the cosmos. Personal efficiency in the *Independent* was in this longer tradition of economy, oversight, and management; in *Engineering Magazine* the word *efficiency* took a more recent form, technical, specialized, and quantified.

Since the mid-nineteenth century technical efficiency had come increasingly to be quantified and, by the early twentieth century, to be measured in percentages. Expressed in percentages, efficiency was a formal and standardized ratio of a machine's productive effect compared with the work expended in producing that effect. It was a formal standard that measured the internal economy of a machine.[8] Efficiency allowed comparisons between machines or systems of widely different design and function and had come to describe the general goal of machine design: to approach as nearly as possible to a perfect correspondence between output and input.[9]

Efficiency's earliest quantified forms dealt with energy and its conversion and transfer, but by early in the twentieth century engineers and mechanics were experimenting with other efficiency calculations. Some very specialized forms proved effervescent, such as a formula to compare ship efficiencies, based on vessel configuration, speed achieved, and amount of power produced, according to which the *Kaiser Wilhelm II* had an efficiency of 293, the *Lusitania* of 257.[10] Several attempts were made to combine a variety of standardized efficiency measures into one—such as thermal, mechanical, and financial efficiencies.[11] More enduring were new arbitrary standards of efficiency, measured in percentages but on scales with arbitrarily fixed points, assigned by management rather than according to physical or mechanical laws. In a railroad shop, for example, 100 percent efficiency might be arbitrarily set at fifteen railroad cars painted per ten-hour shift. Both *Engineering Magazine* and the *Independent* carried examples of the new arbitrary method, and it has become a standard way to set gauges for evaluating efficiency.

Such calculated percentages were not the only way in which engineers approached efficiency. They also discussed it as the foundation of industrial harmony and the goal of good government—efficiency in its more traditional sense. This is evident in two important issues for mechanical engineers that touched on efficiency and reform: the regulation of public utilities and the establishment of a uniform code for the design and production of steam boilers. Debates about utility regulation and the boiler code turned on questions of economy and public management, threats of monopolistic practices, and concerns for public safety and commercial uniformity. Efficiency, in the context of debates over utility regulation and the boiler code, meant a job well and economically done. As Samuel Haber put it, "Efficient and good came closer to meaning the same thing in these years than in any other period of American history."[12]

The treatment of efficiency by *Engineering Magazine* and the *Independent* exemplified many of the characteristics that scholars have considered significant in the Progressive era. Efficiency was an issue of expertise in both journals, in a society increasingly aware of expertise but not certain how much to trust it. Both journals placed efficiency in the context of an increasingly bureaucratized and organized, and thus increasingly impersonal, order of life. They offered efficiency as a measure of reform, in response to the tensions of an industrializing society. And both journals recognized the era as a turbulent one, but the efficiency prescriptions they offered were markedly different. Efficiency was a technical matter in *Engineering Magazine,* offered by experts for use by managers, of projects, departments, or enterprises. Efficiency, in the *Independent,* was also offered by experts but for use by people on themselves.

Technical Efficiency in *Engineering Magazine*

Engineering Magazine was a major source of technical and engineering news. Efficiency appeared often in its pages, where it was usually quantified and often expressed as a percentage. Three particular examples illustrate how efficiency functioned in its pages: a series on internal combustion engines; the appearance of a new standard for measuring efficiency, the arbitrary standard, in studies of the Santa Fe Railroad; and the Gantt chart. Engine efficiencies, particularly thermal and mechanical efficiencies, exemplified balance in the sense of an equation, in which both sides must balance out. At issue were the great conservation laws of nineteenth-century thermodynamics: energy may be transformed but neither created nor destroyed; tracking the conversion of energy in an engine required balancing the accounts on both the input and output ends. The Santa Fe Railroad and the Gantt chart displayed balance of a different kind, not in the precise sense of an equation but in the sense of steadiness and imperturbability, of something regular, stable, and predictable. In all three instances efficiency was an exercise in control.

Engineering Magazine was a witness to the workings of technical efficiency. Directed toward readers who were technically and mathematically trained, it encouraged them to base their social contributions on professionalized status, primarily as mechanical engineers but also as physicists, civil engineers, and, increasingly after 1900, as industrial managers and governmental officials. *Engineering Magazine* came out monthly, each issue compact and dense, sitting heavy in the hand. It was composed of close-copy text, mathematical formulas, and statistical charts and tables, alongside drawings and photographs of instruments, machines, and construction sites. Its reach was international and grounded in advanced formal training, its contributors' names often prefaced by the title "Professor." Between 1907 and 1911 several leaders in the Progressive efficiency movement published the first versions of their seminal works in the *Engineering Magazine*: Harrington Emerson's *Twelve Principles of Efficiency* appeared in serial form from 1909 to 1911, and the magazine was among the first to publish Gantt's influential efficiency charts.

As the main clearinghouse for engineering information in the United States, *Engineering Magazine* was a publishing firm as well as a journal. To the journal the clearinghouse contributed the monthly "Engineering Review" and "Engineering Index," which abstracted nearly two hundred prominent scientific and technical journals published in English, French, German, Dutch, Italian, and Spanish— among them *Deutsche Bauzeitung* from Berlin, the *Far Eastern Review* from Manila, the *Australian Mining Standard* out of Melbourne, and the *Journal* of the South African Association of Engineers, Johannesburg. The "Engineering Index" offered full reprints of indexed articles, in their original languages, for prices ranging from 20¢ to $1.60. Subscribers could also order index entries printed on one side of the paper only, ready to be clipped and affixed to their own index cards.

Engine Efficiencies

Engines were prime vehicles for technical discussions of efficiency, and *Engineering Magazine* ran a series on internal combustion motors from the fall of

1911 to the spring of 1912. It was an auspicious time for engine analysis. In 1911 P. M. Heldt published the first edition of *The Gasoline Motor,* which became a classic text and went through a dozen editions by the 1940s. There were also signs that the treatment of engine efficiencies was becoming standardized, in the increasing use of abbreviations for complex quantities and the assumption that methods of measuring engine performance were well understood. Internal combustion engines had begun to rival steam power in operational costs, and engineers and managers in cities, large businesses, and institutions were looking for information to help them decide which engines they might erect to replace old or failing ones or to provide power in new constructions. Internal combustion was also running individual motorcars, and the *Horseless Age,* which first printed Heldt's engine theory, was as much a motor travel magazine as a technical journal. Internal combustion engines raised theoretical and mechanical issues; they also, as Heldt's publishing in a roadster magazine suggests, stood on contested professional and expert terrain. Engine science drew both from the shop culture of mechanics trained through apprenticeships and from the ascendant practice of engineering education in colleges and universities.[13]

Engineering Magazine's internal combustion series provides an illustration of technical efficiency in several precise forms. It was not a treatise on engine theory but a practical survey of the range of engine options available to engineers, planners, and managers, and it focused on engines already erected and in use. The first article ran in October 1911, opening with a brief history of internal combustion motors and classifying them according to type of fuel used. The bulk of the piece surveyed engines using liquid fuels. Ten photographs of engines, some full-page, four sectional drawings, and four graphs of efficiencies and costs supplemented the text. Subsequent installments analyzed engines using various gas and solid fuels, compared costs of operation for combustion and steam engines, and examined engine design and use in Germany, where some believed that the pressure of scarce fuel resources had stimulated development of the most efficient engines.

Calculating an engine's efficiency was complex, as illustrated in a table showing the performance of an engine installed at Pennsylvania's State Hospital for the Feeble-Minded, in Polk. The table, which appeared in the second installment of *Engineering Magazine's* internal combustion series, illustrates the interplay of theoretical and mechanical factors in measuring engine efficiencies. It shows the results of tests of the engine at loads ranging from one-quarter to full power; the tests determined whether the machine was acceptable to the owner and had met the specifications of the contract and were likely run at the manufacturing site

Date of test, Dec. 8, 1910.

Time	R.P.M.	B.H.P.	Load	Meter reading cu. ft. gas	Std. gas per B.H.P. hour	B.T.U. per B.H.P. hour	Efficiency based on B.H.P. per cent
1-08-05	273	51.8	¼	0	16.9	16,900	15.05
1-18-05	273			146.5			
12-50-05	272			0			
12-55-05	272	100.9	½		12.18	12,180	20.90
1-00-05	272			207.5			
12-36-05	269			0			
12-41-05	269	146.4	¾	122.0	10.02	10,020	27.00
12-46-05	269			245.0			
12-20-05	267			0			
12-25-05	268	199.0	Full	144.0	8.33	8,330	30.60
12-30-05	268			287.5			
11-52-05	267	219.0	10 per cent. over	0	8.21	8,210	31.0
12-02-05	267			307.5			
1-32-05	266		30 per cent over	0			
1-33-05	266	261.0		68	7.62	7,620	33.4
1-34-05	266						

Barometer, 29.5 inches. Gas temperature, 65° F.
Gas pressure at engine, ¾ inch water.
The B.t.u. per cubic foot of gas at working conditions was 943. Corrected to 29.92 inches barometer and 60° F., this became 1,000 B.t.u. per cubic foot. The heating value of the gas was determined by a Junkers calorimeter.
AUTHOR'S NOTE.—The results of these tests should be used with care, as the tests were run for only ten minutes. The results of the last test are obviously wrong as is shown by the efficiency curve, Figure 41. It will be noticed that this test was run only two minutes.

Efficiency of a natural gas engine installed at the Pennsylvania State Hospital in Polk. The final column represents a measure of thermal efficiency. Robert L. Streeter, "The Internal-Combustion Engine in Modern Practice," *Engineering Magazine* 42 (1911): 204. Courtesy of the Minnesota Library Access Center, University of Minnesota; photo by Nancy Johnson.

before the machine was disassembled, shipped, and installed.[14] All but the last test ran for ten minutes, and even ten minutes was short; many engine tests ran much longer. The data columns in the hospital's engine performance chart include time and duration of the test, rotations per minute, brake horsepower (bhp), load, and three columns on the amount of fuel used and its heat value. The far right column shows the efficiency of the engine in brake horsepower percentages; its efficiency increased as load increased. A note beneath the chart indicates that the heat value of the gas was determined using a Junker's calorimeter. The accompanying text specified that it was a natural gas engine of moderate size (200 hp) and named the engineers who had erected and inspected it (J. B. Meriam and George B. Hays, for the hospital and the state, and N. C. Wilson and F. H. Kinol, consultants from Pittsburgh).

The table is interesting both for what it does and does not include. There is no measurement of mechanical efficiency, the term Haber used to indicate output-to-input ratios. There is a measure of thermal efficiency, the quantity that most clearly tied engine analysis to the physics of energy; it appears in the final column on the right, under the complex title "Efficiency based on brake horsepower per cent." The complexity of the column title betrays an intricacy that the simple phrase *thermal efficiency* does not, and in fact mechanics and engineers routinely distinguished between thermal efficiency calculated according to brake horse-power, a measure of output at the drive shaft or other working part, and thermal efficiency measured according to indicated horsepower, a front-end measure taken at the cylinders themselves. They wrote of no single, unified quantity called "thermal efficiency," and calculations of brake horsepower thermal efficiency and indicated horsepower thermal efficiency would differ, even for the same engine.

Such basic measurements as mechanical and thermal efficiency relied upon judgment, manipulation, and interpretation. Determining engine efficiencies required considerable manipulation of the engine and a choice of diagnostic tools, among them power brakes, dynamometers, and calorimeters. Efficiency analysis combined theoretically defined quantities, such as energy, with empirically derived relationships, for example between engine size and power.

Efficiency measurements also differed depending on the end of the process at which they were calculated—at the front end, input, or at the back end, output. Fuels posed a challenging and basic issue in engine efficiencies at the front end. Fuels were not standardized. Machines ran on liquids, kerosene, gasoline, alcohol; oils of various weights; and gases, including illuminating gas derived from bituminous coal and natural gas and gas from anthracite, lignite, peat, wood, and even cinders, collected from locomotive smoke boxes and reused.[15] Calculating engine efficiencies required knowing the heat value of the fuel, typically measured in British thermal units (Btu) per cubic foot; the value varied with barometric pressure and ambient temperature. Many engineers and mechanics measured the heat value of the fuel themselves, as did those who tested the Pennsylvania State Hospital's engine, which is apparent in the note beneath the chart; the American Society of Testing and Materials would later adopt and publish standard tables of fuel heat values, but it remained common for engineers and mechanics to determine fuel heat values themselves.[16]

Fuels posed only one of the input challenges; another important front-end measurement was indicated power, or indicated horsepower, so named because it relied upon an indicator, a device to measure the pressure and volume of the gasses in the cylinder. James Watt had used an indicator to make his famous pressure-volume diagram for steam engines.[17] Internal combustion motors were faster than steam engines and required different sorts of indicators than Watt had used; in fact, the indicators available at the time of the *Engineering Magazine's* series were practically useless above 500 rpm, at which point the inertia of their own parts became great enough to undermine their accuracy. Even for larger and slower engines, indicators were most commonly used in the experimental stages, rather than in the testing of mature, assembled designs.[18]

One of the principal uses of indicated horsepower was in calculating an engine's mechanical efficiency. *Mechanical efficiency* had a precise meaning: how effectively a mechanical engine or system performed, expressed as the ratio between the power an engine actually delivered and the power supplied to it. Necessary losses to friction and heat among its own moving parts accounted for the difference between the two measures. Measuring mechanical efficiency required comparing indicated horsepower, a front-end measurement of power supply, with brake horsepower, a back-end measurement of power delivered. It was also possible to get a good but indirect idea of an engine's mechanical efficiency without using indicated horsepower, in a process that came to be called "motoring": it involved attaching the engine to another motor, an electric motor or a dynamometer, and using the second motor to turn over and run the first. The power it took would give a fairly good measure of the first engine's frictional losses, and from that the engine's mechanical efficiency could be calculated.[19] There is no evidence in the table that either indicated horsepower or mechanical efficiency was calculated for the Pennsylvania State Hospital engine; by this time a well-built engine was generally assumed to have a mechanical efficiency of between 80 and 90 percent.[20]

Much more commonly assessed than indicated horsepower was brake horsepower, a fundamental value measured for nearly every engine. It was a direct and physical assessment of the power of the engine embodied in the torque, or friction, on its crankshaft, flywheel, or other fundamental moving part, as measured by a dynamometer or power brake. This, the brake horsepower, along with the amount of fuel used and its heat value, were the important measurements made on the Pennsylvania State Hospital's engine. Both brake horsepower and indicated horsepower were direct physical measurements of a machine in motion and depended on devices attached to the engine itself, to its cylinders or working parts. Measuring the amount of fuel used, and assessing its heat value, depended also on a variety of instruments. Assessing an engine's mechanical or thermal efficiency did depend on the fundamental law of the conservation of energy, which was itself an elegant and clean physical concept, but determining how that energy was distributed, or to which accounts it should be charged, was messy and complex.

The efficiency of the Pennsylvania State Hospital's engine, measured in the final column of the table, was one of balance. It brought into equilibrium the equation of energy out and energy in, through whatever devices they were measured. The percentage in the final column was derived from a fundamental relationship between energy, matter in motion, and heat, a relationship embedded in both the third column, brake horsepower, and the seventh, Btu per brake horsepower hour. The equation quite literally balanced: the percentage in the final column, multiplied by the power expended (energy in), would yield, or equal, the power produced (energy out). Units would have to be converted, most notably Btu and brake horsepower per hour, but the accounts would balance. They had to, in this instance, because the efficiency calculation was governed by a theoretical maximum: as a matter of theory, the engine could not deliver more energy than it used up, whether as heat, work, or friction. Energy in and energy out were balanced and accounted for in the efficiency percentage.

In the case of engines, balanced efficiency provided a generalizable analytical tool in the face of enormous variety—engines of different sizes, using different fuels, doing different work. Beneath this balanced result was a changing and contingent system of measurements using a variety of instruments. A range of efficiencies could be calculated for any given engine, and even engines of the same make might display differences in efficiency.

Arbitrary Standards

Engineers also began to use percentages in arbitrary measurements, in which efficiency was governed not by a theoretical maximum, such as energy, but by an arbitrarily chosen standard. With arbitrary standards it became possible to speak of, and measure, efficiency that exceeded 100 percent. A number of engine efficiencies used arbitrary standards—weight efficiency, for example, and fuel efficiency—but arbitrary standards were particularly important to engineers and consultants associated with new management techniques.

Walter Polakov, a prominent member of the circle of management consultants deeply influenced by Taylor, was one of the first to describe the arbitrary nature of newer measures of efficiency. He recognized, writing in *Engineering Magazine,* that it was not always possible to use energy measures and suggested that acceptable standards could be created, using instead what Harrington Emerson called "equivalents," or commensurate units of measure that eventually canceled each other out. This was akin to setting the gradations on a thermometer or using sea level as the basis for drawing railroad profiles, and Polakov offered an illustration: assuming as a baseline that 1 million Btu of coal could be purchased for five cents, purchasing 1 million Btu at 6 cents would yield an efficiency of 83.5 percent ($\frac{5}{6}$). Quite simply, as Polakov put it, fixed points in an arbitrary standard of efficiency were to be "chosen arbitrarily."[21]

Polakov attributed the method to Harrington Emerson, who had advocated the use of arbitrarily chosen efficiency standards in cases in which ideal or theoretical efficiency could not be determined. He used the yield of potatoes in the United States as an example: "The average yield per acre over a series of years is 96 bushels. Shall we therefore set 100 bushels as standard 100 per cent efficiency?" Emerson chose this example because he believed management to be the important factor, not soil or climate; after all, the "desert state of Wyoming" had produced the potato efficiency champion in 1907, with an efficiency of 200 percent with a yield of 200 bushels per acre.[22]

The element of balance in arbitrary standards lay in their emphasis on predictability. When theoretical limits could not be found, arbitrary goals were to be established systematically and in advance, or current performance and future improvements could not be assessed. "Predetermination of results is the main characteristic of the modern method," Emerson wrote; the purpose, from the manager's point of view, was steady, reliable, unperturbed output.[23] This was balance less in the sense of an equation than in the sense of a thing unwavering, well regulated, and controlled. Control was central to arbitrary standards; the choice of scale reflected what managers believed could be controlled, and the gradations on the scale represented levels of control. Emerson's unit of choice was a "standard time for a unit operation," with efficiency measured on a scale from zero to one.[24]

The arbitrary efficiency percentage had real management uses at a time when management tools were important to engineers because they offered an elite and expert status in modern corporate management, even as opportunities for shop leadership waned.[25] Emerson had made a name for himself by deploying and creating management tools in reorganizing the Santa Fe Railroad's machine shops, centrally important because they maintained and repaired the Santa Fe's rolling stock, including its locomotives. His principal tools were standardized tasks, completion schedules, and arbitrary efficiency standards; between 1904 and 1907, under his reforms, machine shop productivity increased by 57 percent. The Santa Fe, along with other railroads, faced declining rates per ton- and passenger-mile and increasing expenditures due to rising standards of service. Emerson's Santa Fe system would later be subjected to harsh criticism, especially in light of his testimony in the Eastern Rates case, but it was initially greeted with respect and enthusiasm, and recent historical work suggests Emerson's reforms were important and enduring. *American Engineer and Railroad Journal* and the *Railway Age Gazette* sent editors to inspect Emerson's reforms, and both published favorable reports.[26] *Engineering Magazine* sent its editor, Charles Buxton Going, and published his observations in a series running from 1909 to 1910. Not only the railroad's profitable performance but also its scale of operations had attracted wide attention. It served cities as far apart as Kansas City, Albuquerque, Galveston, Los Angeles, San Francisco, and others up the Pacific Coast and had to contend with extraordinary technical challenges: steep grades, mountain passes of nearly eight thousand feet, desert conditions, and, in New Mexico near Raton, a combination of all three.

Under Emerson's direction the Santa Fe's management had chosen a method forcing costs and compensation "into parallel," using time spent at task as the basis for evaluating not only the performance of laborers but also of engines.[27] In the centralized machine shops, equipped to produce and maintain all facets of the system, parts, tools, and operations were standardized and operations described in precise language and at length in order to eliminate the need for decision or judgment on the part of a worker. Coach Repair Schedule Number 0228, for example, allowed two hours of "one-man time" for "doors, sliding, on baggage cars or doors of similar construction, [to be] scraped and dressed after paint has been removed." Committees drew up work and task schedules with careful attention to standardized wording, to make "immediately apparent" any variations or defects in methods, practice, or machines. Many engineers claimed that efficiency measures did not suppress individual talent and initiative, as critics had argued, but rather made more evident those workers of exceptional gifts. Emerson was among them; he wrote that the unit of man-hours made "the efficient men as to steadiness and output loom up like mountains from a plain and the inefficient appear like sink-holes."[28]

The control exerted by arbitrary standards was revealed in the Santa Fe's bonus system of wages and in the efficiency education given to apprentices. The calculation of bonuses required an enormous administrative effort, with all tasks measured in man-hour units and the efficiency of each worker assessed. Mechanics filled out an "Individual Effort Check" for each standardized task they performed, on which they could themselves calculate their efficiency and the bonuses to which they were entitled.[29] The efficiency percentage was reinforced in apprenticeship, during which new mechanics learned how to apply efficiency standards to themselves and to their machines. Many of the problems assigned to apprentices on the question of percentages dealt with measuring efficiency and then calculating bonuses, asking, for example, what was the "per cent efficiency" of "a second year apprentice at Cleburne [who had] accomplished in 249.7 hours an amount of work for which 283.8 hours standard time was allowed . . . If his ordinary wages amounted to $39.95, how much bonus did he receive?"[30]

The Santa Fe system illustrated the interconnection of efficiencies of balance and management success, not only in the form of increased profits but also in the form of control. Going had observed an informal and modified "military discipline and control" asserted over apprentices, both in the shop and in public, and worries about losing control, specifically over disorder in the shops and strike threats, had motivated the Santa Fe to hire Emerson in the first place. One important result of Emerson's work was that it had helped to keep the Santa Fe's shops union-free. The efficiency Emerson strove to create, and which Going described, was an efficiency of balance—steady, reliable, and regulated. Efficient workers were productive because they were steady, dependable, and followed the schedule of tasks, not because they were exceptionally speedy or skilled. Efficient workers provided a steady and reliable platform that managers might use to generate savings, profits, or growth.

Gantt's Chart

Gantt's chart visibly demonstrated an efficiency of balance. Efficiency, as Henry Gantt put it, meant equilibrium and stability, and his charts were intended primarily to stabilize and regulate production.[31] Such charts were Gantt's contribution to an ongoing debate over how workers should be paid, specifically over whether they should receive a piece rate, be allowed bonuses, or be paid for their time. A piece rate, in which workers were paid a set amount for each piece they produced, at first motivated workers to greater productivity, Gantt noted. The problem came when workers started consistently producing more and their managers responded by cutting the rates per piece. Workers then had to produce still more in order to break even. Because workers could foresee such rate cuts, they were not motivated by the piece rate system, as they might otherwise have been. Gantt proposed what he called a differential piece rate, in which workers who met a daily quota received a higher rate for each piece; he aimed not only to stimulate production but also to make it predictable. Managers would then know what they could expect, and workers could count on a certain amount of pay.

The chart of the young women in the bleachery was a daily record of whether they had earned the bonus rate under Gantt's system. Workers who had agreed beforehand to attempt to meet the quota, and succeeded exactly, received a bonus rate per piece. The bonus rewarded 100 percent efficiency, measured against an arbitrary standard that set 100 percent equal to the previous performance of one or more outstanding workers. The chart was laid out as a grid, dates in the columns running along the top, the names of the workers running in a vertical column down the left. Each box on the grid represented one worker's performance for one day, and the boxes were filled with black or red marks, depending on that performance. Only efficiency percentages of 100 or zero were assigned or recorded. Solid black filled the daily boxes for workers who had earned the bonus on that date. Workers who did not attempt the bonus received a black *X* and the usual day rate. Workers who were absent received a red *X* (here shown as gray), and workers who attempted to earn the bonus but failed received solid red boxes. Gantt called the amount of black "a measure of efficiency."[32] The chart measured not only the efficiency of the women but also that of their managers. The more black boxes on the chart, the more efficient and reliable the women's production was, and the better their manager's efficiency; red meant failure, a worker's failure to meet a predetermined quota and a manager's failure to ensure uniform productivity. Red not only meant a worker was inefficient; it also indicated inefficient management.

Gantt was concerned not only with speed and increased production but also with achieving and maintaining control over the internal economy of production. The more stable the stream of black marks on the chart, the better. A red box by a worker's name did not indicate poor performance, for a laborer could produce more than he or she had under the day wage rate but still not meet a newly specified quota. Workers were rewarded, in other words, for consistently and predictably meeting a goal established ahead of time. The balanced efficiency of the young women, stable and predictable and reflected in the increasing amount of black on the chart, was the foundation upon which a manager might seek to achieve additional, often transformative goals.

Technical efficiency was complex. Even in what Walter Polakov called its unassailable form, when, as in engines, it was tied to energy measurements, calculating efficiency involved manipulating instruments and machines and making choices about what types to measure and which tools to use. Gantt and the Santa Fe Railroad developed complex tools to quantify efficiency, including percentages and all-or-nothing charts, using arbitrary baselines such as man-hours and full-bonus black boxes. Overlaying the complexity of these measurements was an emphasis on regularity and stability. Despite the complexity of their various measurements, the efficiencies here described were ones of balance. This was most apparent in engine efficiencies, in which the efficiency percentage quite literally brought into balance the accounts for energy out and energy in. In the Santa Fe's machine shops and in Gantt's work balance appeared in the emphasis on stability, regularity, and equilibrium. That balance, that stability, required control.

Personal Efficiency in the *Independent*

The *Independent* emphasized personal efficiency, to be achieved through self-control. Its editors had worried that the benefits of industrial efficiency accrued to owners at a much greater rate than to workers; they had also applauded efforts to bring efficiency and transparency to government. But the journal truly invested itself in personal efficiency. From 1914 to 1917 it sponsored a biweekly column of efficiency advice and offered a mail-in efficiency service, to which readers could turn for efficiency news and aids. The *Independent* also inaugurated a series of feature articles on personal efficiency, the only series to appear in its feature section. Personal efficiency in the *Independent* relied upon control, particularly self-control, and was aimed at regularizing and stabilizing behavior. Control was most apparent in a series of quizzes that accompanied the efficiency features; the balance at which personal efficiency aimed was apparent also in the queries readers sent to the journal's efficiency service and the answers published in its efficiency advice column, "The Efficiency Question Box." Through self-control personal efficiency offered a way to manage the turbulence and friction of a changing time, as technical efficiency had offered a way to manage turbulence and unsteadiness in machines and workers. Personal efficiency, too, aimed at balance. But an important distinction set discussions of efficiency in the *Independent* apart from those in *Engineering Magazine:* in the *Independent* personal efficiency was embedded in a rhetoric of efficiency's larger and dynamic transforming power.

The *Independent* was "a journal of genuine importance," Frank Luther Mott observed in his monumental survey of the history of American magazines. The journal was a long-standing organ of reform, founded in 1848 as a religious weekly by Congregationalists outside of New England who wanted their own journal of opinion and news. Opposing slavery was a second and serious purpose, and for some time the journal's managing editor was Joshua Leavitt, Free Soil Party organizer and former editor of the *Emancipator.* The brilliant antislavery writer and preacher Henry Ward Beecher contributed articles and opinion pieces and later served as editor. By the turn of the century the *Independent* had given up its religious emphasis and, although increasingly nonpartisan, did advocate many reforms, including the breakup of monopolies and ameliorating conditions for industrial workers.[33]

By 1910 the *Independent* had become a weekly for general readers, and it was slim and light in the hand. In 1913 it adopted a large format, nearly tabloid size, and began increasingly to use illustrations and photographs, including formal portraits of leaders in business, politics, and the arts and shots of news in the making. The *Independent* aimed at well-educated readers, many in positions of relative social privilege, including professionals, and those who wished to become better educated and take up socially useful positions: women who believed their talents lay beyond family pursuits, students eager to prepare themselves for a new international milieu, and workers seeking routes of advancement and social contribution. The journal was a long-standing advocate of personal efficiency and stayed with the issue longer than other periodicals; it sponsored symposia on efficiency in addition to its question-and-answer service and biweekly column.[34]

The *Independent's* first "Efficiency Number," November 30, 1914, illustrated the qualities for which the *Independent* was best known: serious opinions on matters of public interest, here on Wilson's new system of Federal Reserve banks and the implications of Germany's war aims; support for reform, instanced in the editors' enthusiasm for the appointment of Thomas Mott Osborne, chairman of the New York Commission on Prison Reform, as the new warden of Sing Sing; and assistance to readers interested in self-improvement, apparent in the efficiency advice it offered and attested to by an abundance of advertisements for self-study. The centerpiece of the issue was a photograph of the sinking of the super-dreadnought HMS *Audacious* off the coast of Ireland on October 30, 1914. Several photos of Belgian troops appeared: riflemen fighting near Ypres, a Red Cross nurse visiting a trench near Nieuport, and a boy helping to bury Belgian soldiers in a ravaged church graveyard at Pervyse. The back pages were filled with columns, including book reviews and discussions of the markets and insurance, and with advertisements—for books; business equipment including desks, cabinets, letter openers, and typewriters; personal goods such as watches, boots, flowers, cigars, and furs; and courses of study, in law, shorthand, and public speaking. A page of humorous comments by literary editor Edwin Slosson closed the issue; he suggested that Henry Morgenthau be made sultan of Turkey and noted that the new *English Dialect Dictionary* had only thirteen hundred names to call a fool: "Still the supply is not equal to the demand."

Two pieces on efficiency took up most of the space devoted to feature articles.

The first, "What Is Efficiency?" an introduction to efficiency by popular author and consultant Edward Earle Purinton, included a personal efficiency test. The second was a symposium on "What Efficiency Means to Ten Efficient Men" and included statements by Louis Brandeis and Harrington Emerson, along with Charles Eliot, president emeritus of Harvard; New York City mayor John Purroy Mitchel; and Judge Elbert H. Gary, chairman of U.S. Steel.

The *Independent* had been interested in efficiency for some time, and its initial editorial stance was skeptical. Although they were in favor of efficiency when presented as a technique to bring transparency and responsibility to government, the editors worried that industrial managers and owners would employ other sorts of efficiency to increase their returns at the expense of their workers. An editorial shortly after the Eastern Rates case charged that efficiency rewarded employers at a rate "3–10x greater" than it did workers and might well exacerbate a gulf between workers and employers. The journal later indicted efficiency as creating "mechanicalized life." In these concerns the *Independent* reflected widespread worry about the character of the nation's developing industrial society and the social costs and personal unease many people felt as they left old forms of life characterized by personal and local connections, and largely rural, to join the more impersonal social structures of the United States's growing cities and newer and larger factories, which were administered often without regard for personal welfare. Many Americans felt vulnerable, and some experienced genuine personal trauma.[35]

While efficiency measures imposed on workers by management worried the editors, the journal wholeheartedly endorsed personal efficiencies that people could practice on themselves. In 1914 the *Independent* kicked off its series "Efficiency and Life" with Purinton's article "What Is Efficiency?" Purinton's efficiency credentials lay in his popularity as an author of inspirational guides to daily life. He had written the widely known *Triumph of the Man Who Acts* (with more than 700,000 copies in circulation); the Pennsylvania Railroad, the New York Department of Education, the National Cash Register Company, the Wanamaker stores, and the United States Army had distributed copies. Purinton's other titles included *Lords of Ourselves* and *A Daily Guide to Efficiency;* he would later rework his *Independent* articles into a book entitled *Efficient Living.*[36]

In his first piece for the *Independent* Purinton employed a language of dynamic transformation. He connected a balanced efficiency of self-discipline and personal stability to a dynamic and transformative efficiency of power. At its foundation efficiency was "the science of self-management," he wrote; this self-control was the root of all other efficiencies and of true social progress. "Efficiency is more than speed and economy—it is the reeducation and reconstruction of men," he asserted, echoing Taylor, who had described scientific management as a mental revolution.[37] Purinton provided motivation for the rigors of self-discipline in a rhetoric of triumph. "Efficiency is the difference between wealth and poverty, fame and obscurity, power and weakness, health and disease, growth and death, hope and despair. Efficiency makes kings of us all. Only efficiency conquers fate . . . Efficiency leads us from a world of chance to a realm of choice, changing us from automatons to men. Efficiency provides our only freedom—that of shaping circumstances and hewing events to suit ourselves!" Purinton encouraged his readers to "learn conservation and direction, thru efficiency. Then we shall rule the world."[38]

This language provided an ambiguous context for the efficiency measures offered in ensuing *Independent* features. Efficiency was portrayed as a way to power and wealth, yet readers were counseled to make themselves into good employees. Purinton's first article came in the early days of the European war, just months after Austria had bombarded Belgrade, and although he seldom mentioned the war itself, he did write of efficiency as the basis of world supremacy, mastery, and rule. The transformative efficiency of wealth, power, hope, and rule would be founded on self-mastery, but the marks of that self-mastery were balanced and conservative, not transformative—stability, regularity, reliability. This suggests that there was an alternate form of efficiency to that of balance, a transformative, dynamic form.

The *Independent*'s efficiency series was well received. The *Ambassador,* of Lockport, New York, reprinted several paragraphs of the first installment, its editor calling it "the best article on efficiency I have ever read."[39] A reverend from Wisconsin wrote praising the "vital gospel" Purinton preached,[40] and the office manager of a small business in Vermont called Purinton's article "exactly the spur" he needed to accomplish the most his position allowed.[41]

Measurement was central to personal efficiency. Each article in Purinton's efficiency series included an exercise, a chart or table, designed to reinforce the efficiency lesson; many of them were tests measuring efficiency against an arbitrary scale with a maximum score of 100 percent.[42] These tests employed the efficiency percentage in its technical form: they required close surveillance of a person's internal and personal workings, accounted for them in commensurate units, in this case arbitrarily assigned point values, and yielded percentages as a result. Tests measured a variety of personal efficiencies, including efficiency in the home, in recreation, in money management, and in eating and nutrition.

The first such test, the "Personal Efficiency Test," included thirty questions.

PERSONAL EFFICIENCY TEST

DIRECTIONS. In answering questions write 100 for "Yes," 0 for "No." If the answer is a partial affirmative write the number between 0 and 100 that expresses the degree of assurance. Then add the column of percentages, divide the total by 30, and the answer will be your approximate grade in efficiency. The value of the test lies in the honesty of the answers.

ANSWERS (in per cent)

1. Do you like your work?
2. Have you learned the best, quickest and easiest way of doing it?
3. Are you thoroly informed on "scientific management?"
4. Do you know where your greatest power lies?
5. Have you a fixed goal, in line with your supreme talent?
6. Do you believe absolutely in your own future?
7. Are you in perfect physical health?
8. Have you learned how to get well and keep well?
9. Can you be optimistic, under all circumstances?
10. Do you realize which of your habits, thoughts or emotions make you inefficient?
11. Have you made an inventory of your mental and moral traits?
12. Are you correcting your known weaknesses; mental, financial, social or spiritual?
13. Have you discovered which foods, baths and exercizes increase your energy and heighten your mentality?
14. Do you breathe deeply and hold an erect posture?
15. Is your sleep long and dreamless and refreshing, with your sleeping-room perfectly ventilated?
16. Do you drink three pints of pure water daily?
17. Do you eat slowly, moderately, regularly?
18. Is all your clothing made loose, to allow blood and nerves free play?
19. Are you independent, fearless, positive?
20. Are you tactful, cautious, courteous?
21. Have you secured the best possible advisers and associates?
22. Are all your co-workers eager to help make your plans a success?
23. Do you wish your rivals well, and never speak ill of them?
24. Do you work harder than anybody else in the business?
25. Have you learned the science of planning your day ahead?
26. Can you relax entirely in your leisure hours?
27. Are you saving money systematically?
28. Do you enjoy art, music, literature, and the presence of little children?
29. Does your highest ambition include some real service to Humanity?
30. Have you a great love in your life, to steady, cheer and empower you?

Divide the total by 30

NOTE. A complete Efficiency Test would include other vital questions, but answers to these will furnish a self-analysis of approximate reliability.

The quotient shows your percentage of efficiency

Copyright, 1914, by Edward Earle Purinton

Personal efficiency test. Calculated as directed, the score would be a percentage. Edward Earle Purinton, "What Is Efficiency?" *Independent* 80 (1914): 323. Courtesy of Wilson Library Annex, University of Minnesota; photo by Nancy Johnson.

Readers were to give themselves a score of between zero and 100 for each and then divide the total by 30 to get a percentage. Questions included "Do you like your work?" and "Have you a great love in your life, to steady, cheer and empower you?" Readers got points for having perfect physical health; for being independent, fearless, tactful, and courteous; for enjoying "art, music, literature, and the presence of little children"; and for wanting to be of "real service to Humanity." Optimism and self-correction also helped the score. Purinton judged that the average grade on the personal efficiency test was 40 percent.[43]

The *Independent's* personal efficiency tests used arbitrary standards yielding a maximum score of 100. Many of them took the form of questions worth varying points, to be added up for the total score; others were made up of questions of equal weight, with the final tally to be divided by the number of questions. A score of 80 percent on any of these quizzes was extremely good, Purinton wrote, and noted that most people scored between 30 and 40 percent.

Its mix of reasonable and quirky advice underscored the arbitrary nature of the efficiency tests. The "Efficiency Money Quiz" included the reasonable question of whether one would "cheerfully go without luxuries" in order to save and a provocative one suggesting that one should resolve to "amass a competency" that would allow one to do "something in life more valuable than making money."[44] The "Efficiency Food Quiz" asked whether one enjoyed meals, liked fasting "as well as feasting," thoroughly masticated one's food, and had trained the palate "to know and reject wrong combinations, e.g., pickles and milk."[45] The "Health Efficiency Gauge" asked questions about freedom from pain and fear of disease, sanitary appliances in the home, and "knowledge of mental and spiritual factors in health" and suggested that a higher percentage of "health status" could be had by daily exercise in the open air and "enjoyment of same."[46]

Control, specifically self-control, was the point of the tests. They encouraged continual personal surveillance and adjustment and illustrate what Robert H. Wiebe has called the "techniques of constant watchfulness and mechanisms of continuous management."[47] These were techniques of self-management, as technical efficiency was a tool of industrial management.

Personal efficiency in the *Independent* exhibited not only elements of control but also a goal of balance. That balanced quality is most apparent in the paper's "Efficiency Question Box," its biweekly column answering readers' efficiency queries. The questions readers sent in reveal how deeply efficiency had penetrated personal life, and the responses, often from Purinton, illustrate how much personal efficiency was dedicated to gaining stability and regularity and how much it was a matter of solid retrenchment rather than transformation. Purinton, in the "Efficiency Question Box," as in his books and articles, was genial and kind, offering realistic suggestions on career training and household management and comfort to people struggling with ill health or feeling lost in the city. The efficiency response to almost any situation, he wrote, was to seek expert advice, recognize that the situation was largely of one's own making, and then stick with it, especially if it was a job.

Readers' questions represented all aspects of life. Miss M.A. of New York wondered how she might cultivate mental poise in a stressful situation and avoid nervous breakdown; she was advised to write to the Goodyear Book Concern of New York City for a list of publications on mind control. A college student from California asked for "any quick guaranteed method" to develop his biceps (Purinton told him that a student of efficiency did not have the time to "bother much with the biceps" and should turn to the muscles surrounding his heart, lungs, stomach, liver, and solar plexus). Mr. R.E.K. of Wyoming, a civil engineer, asked where he might send for a mail course in stenography, and Mr. G.L.B. of Pennsylvania worried because he would become so engaged in mental work that he skipped his meals. "How," he asked, "can I control myself, for better service to humanity?" A young woman, Miss R.Y. of New York, wrote looking for a "morally safe," hygienic, and comfortable place to live on a modest salary in New York City and was referred to the YWCA on West 39th Street.[48]

Seek expert advice: that was uniformly the *Independent's* response to correspondents seeking efficiency counsel. Mrs. R.B.S. of Virginia, interested in promoting the study of efficiency for herself, her husband, and her sons, found herself challenged to read *Good Housekeeping, Ladies Home Journal,* and *American Club Woman.* A New York businessman worried about competition from smaller firms was advised to hire an efficiency expert to discover the "leak" for him or, "on reading the right books," he might discover it for himself. Mrs. M.S.S. of New York asked whether parents were responsible for the "noble or evil careers of their children." Purinton replied that only two simple things were required of parents—first, that they love each other, and, second, that they employ all known scientific and religious methods for raising and training children. Purinton's recommended study list included "thoro study of sex-conservation, prenatal influence, physical, mental, and spiritual aides to motherhood, privileges and responsibilities of fatherhood, psychology and hygiene of babyhood, principles and methods of Froebel, Horace Mann, and other true educators—in short, all that pertains to the growth of a child."[49]

The "Efficiency Question Box" advocated the efficiency of a stable, disciplined, and reliable employee. An office boy would be "a combination of Solomon, Darwin, Caesar, and Christopher Columbus" if he became regular, thorough, and dependable by following Purinton's efficiency work schedule.[50] Purinton discouraged Mr. C.P. of New Jersey, who sought a correspondence course in calculus, replying that calculus was beyond the pale of the person seeking efficiency, being "so theoretical and abstruse" that it was unprofitable to study or teach.[51] "A real man is always a machine while on the job," Purinton wrote. The "recipe for efficiency: be a plodder by day and a poet by night." Whatever plans and dreams a man had, he was to save them for moments when he was alone, striving at all other times to remain ordinary.[52]

This was balanced efficiency. It was stable and well regulated, and it was reliable. Mr. J.S. of New York had, after years of careful attention to his efficiency, been passed over for an expected promotion that went instead to an incompetent relative of the supervisor. How could an employer be made "efficient in justice, conscience, and fair play?" he asked. "What becomes of the years of sacrifice and devotion to the interests of such an employer?" Purinton exhorted him to stay where he was, to "stick, brother, stick," and wait for management to admit its error. Efficiency would triumph in the end, and the deed of injustice would crumble like a house of straw before the first gale.[53] Others wrote asking how to handle themselves in jobs in which efficiency "avail[ed] nothing." Purinton's advice was uniform. In such a situation the employee should first evaluate his or her own failings, remember to be thankful for steady employment, and summon a little grit. If the situation truly seemed hopeless, an employee could begin training for another position during off-hours, as long as it did not interfere with his or her obligations at work. "Your first duty," Purinton repeated, "is to be the finest possible worker in your present situation."[54]

This advice, offered to "Question Box" petitioners, contrasted with the dynamic rhetoric accompanying the *Independent*'s personal efficiency tests in the series of feature articles. Purinton created a vision of dynamic transformation, but the tools he offered were conservative ones, of balance and stasis. The *Independent* described personal efficiency both as the steady foundation upon which great and transforming accomplishments might be built and as the transforming accomplishment itself. It connected the two only through the idea of personal transformation. How personal transformation was to yield world mastery or make one able to hew events to suit one's purposes remained unclear. The journal provided no instructions on how to achieve transformation beyond oneself. Despite a rhetoric of dynamic transformation, the personal efficiency of the *Independent* remained an efficiency of balance. It emphasized stability, reliability, and imperturbability. Self-control, through self-analysis and self-surveillance, would make one manageable, even by oneself.

Interest in efficiency was at its height during the Progressive era in the United States, and examples from this period illustrate how the concept was remade from an intellectual construction into a workable tool. In some instances it was clearly concerned with mechanics and machines, in others with intimate private behavior. Efficiency's multiformity attests to its power, demonstrating its adaptability and the resilience of the notions at its core. Basic congruencies between these efficiencies of apparently quite different kinds, the technical efficiency of an engineering periodical and the personal efficiency of a general interest journal of reform, did appear in *Engineering Magazine* and the *Independent*. Both were efficiencies of balance, aimed at uniting a variety of complex and shifting components into a reliable, stable, and regular platform. Both used tools of control and of a particular sort—standards that simultaneously defined efficiency and served as gauges against which to measure how well it had been achieved. Some of these standards were linked to natural quantities and governed by natural law, as in the case of energy and engine efficiencies; others were assigned arbitrarily, by managers or efficiency experts. Both technical and personal efficiency relied on continual surveillance and adjustment.

Such similarities between technical and personal efficiency suggest it is time to rethink the efficiency dichotomy, between engineering efficiency and efficiency in other forms, that runs through the literature on the Progressive era. The similarities suggest that efficiency, even outside of technical contexts, was more than an engineering metaphor or slogan. The *Independent* was unique in offering a way to quantify personal efficiency, especially in using techniques mirroring engineering efficiency measures, but the constant scrutiny that personal efficiency required, and the conservative message that efficiency meant steadiness and reliability, carried beyond the efficiency quizzes into readers' efficiency questions and the journal's response. This kind of efficiency, an efficiency of balance, aimed not at dramatic increases in productivity—although it was hoped that output would rise—but at guaranteeing a reliable, regular rate of production and cultivating reliable, steady habits of character. It was a tool of self-management and personal stability in the face of turbulent change.

Efficiency was nonetheless embedded in a rhetoric of its dynamic, transformative power. Balanced efficiencies provided the reliable elements of economic or social transformation, the interchangeable and standardized parts, the unchanging substrata, upon which a new bureaucratic order of interaction and adjustment, of change, might be built.

CHAPTER FIVE

An Island of Mechanical Predictability

Efficient Worker Seating in Late Weimar Germany

The logo left no doubt: Siemens had contributed the display of work stations to the exhibit on efficient worker seating and work tables. The chairs were Elmo work stools, developed in an effort to improve worker efficiency at the Elmowerk, the small electrical motors division of Siemens-Schuckertwerke, Siemens's heavy-current company. The exhibit, which opened in Berlin in 1929 and toured Germany during 1931 and 1932, displayed ways to increase workers' comfort and support. It also illustrated an industrial need for regularity and uniformity in human motion and signaled the creation of a pathology of human movements that did not fit regular and uniform patterns. The exhibit and related educational materials demonstrate how an industrial norm of efficiency, which required straight lines and smooth curves of motion in machines, was paralleled in a pathology of human motions. The exhibit offered a remedy: people could be made efficient by a mechanical form of discipline, their movements constrained much as were those of a machine.

The exhibit was part of Germany's rationalization movement, especially prominent in the 1920s, which emphasized industrial standardization and the elimination of waste. A characteristic feature of rationalization was its emphasis on the smooth flow of production, which promised more goods produced more cheaply and thus more available to people lower on the economic ladder. Production, smooth and unimpeded, was intended to address poverty and want and to reshape a divided society into a cooperative and harmonious one. Rationalization might require some new machines, but it would not generally require vast outlays of capital: it would be achieved, instead, by eliminating impediments that caused turbulence and waste. Chief among the causes of turbulence were people themselves, or more precisely human factors, who became increasingly important to the movement during the deepening social and economic crisis late in the

Display that Siemens contributed to an exhibit on worker seating and work tables.
C. W. Drescher, "Arbeitssitz und Arbeitstisch, Ein Beitrag zur Entwicklung," *Siemens Jahrbuch 1930* (Berlin: Siemens, 1930), 439. Courtesy of the Staatsbibliothek, Berlin.

decade. People disturbed flow not only through obstinacy and willful obstruction but in the very way they moved. Even involuntary movements, brought on by fatigue, disrupted flow and required a remedy.

The exhibit provided one, developed by industrial consultants and sponsored by Germany's most prestigious institution of rationalization, the Reichskuratorium für Wirtschaftlichkeit (RKW, or National Bureau for Economy and Efficiency). Ludwig Ascher, director of the medical division of the Social Hygiene Research Institute in Frankfurt, RKW author, and consultant to Siemens-Schuckertwerke, and his colleague Hermann Hebestreit did much of the research on which the exhibit was based; they described an efficiency of static work, in which only certain parts of a worker's body moved while the remainder was held at rest. Workers, they concluded, were most efficient when they sat still.

The RKW mounted the work seating exhibit even as its larger project, intended to improve efficiency by the highest levels of management, failed in a crisis of mass unemployment and social disintegration. The exhibit illustrates the contrast between a static efficiency of balance, aimed at instilling discipline and stability in the workforce, and a dynamic efficiency of management that had already tried, and failed, to transform German society.

Inefficiency and the Pathology of Unbalanced Human Motion

Ascher and Hebestreit sought an efficiency of balance in human movement in which expenditures of human energy would be offset by immediately productive motion. Inefficient motion was unbalanced, and they characterized it as wasteful because it did not produce anything. A concept of static work was particularly important for Hebestreit, who determined that making workers more efficient was not only about teaching them new ways to move but also about keeping them still.[1]

How people move is significant not only in industrial settings; it also has personal, social, and cultural meaning. It reveals gender traits, furnishes evidence of disease, and reflects internalized forms of social or political discipline. Particular medical pathologies, such as hysteria and epilepsy, had long been linked with how people moved, and industrial occupations could also lead to deformities that deranged movement: in a condition called "hindleg," lathe workers developed crooked backs and one permanently bent leg, so together their legs had "the form of a K," and silk twisters might develop a peculiar listing gait, their bodies permanently bent toward the side of the arm that held the heavy skein.[2] How people move their bodies speaks also of their culture: whether they hold their eyes open when swimming or how they run or use a shovel. Such "techniques of the body" are not individual matters but social, acquired through education and mimicry. Synchronized military drills use rhythmic and coordination motions to create group solidarity and "muscular bonding" among soldiers, and some Weimar sex reformers believed that smooth and rhythmic human motion not only distinguished a good worker from a poor one but also set off good from poor erotic performance: a worker of unrhythmic and irregular habits, as one reformer put it, also tended "to practice unrhythmic, inconsistent, jerky coital motions."[3] Some also viewed motion as an important component in eugenic definitions of health and disease: Sheila Faith Weiss has described a relationship between eugenics and notions of national efficiency, and Weimar eugenicists linked eugenics and efficiency to adaptation, arguing that the ability to adapt distinguished both healthy bodies and healthy societies from those that were diseased. Ability to move increased a body's ability to adapt, and it was the ability to adapt that distinguished a living body from a dead one. The dead body did not adapt; it did not move.[4]

Standardizing human motion was thus a complex problem with personal and social implications, and Ascher himself referred to the human organism as "a living protest against the norm." Applying norms to machines was relatively easy, he argued; applying them to people was more difficult. People possessed will and had lots of moving parts, which meant that they could easily adapt but also that they could protest. Ascher's work illustrates how an industrial norm of efficiency, which required straight lines and smooth curves of motion in machines, was paralleled in a pathology of human motions. As Ascher saw it, workers' bodies embodied not the smooth and controlled motion of the norm but its inverse, a pathology of jerking, shaky, or indiscriminate movement.[5]

Ascher and Hebestreit created the pathology of unbalanced and inefficient human motion by comparing how workers did move with how they should move, measuring their movements against a developing norm of efficient human motion patterned along industrial lines. Georges Canguilhem, in his influential study of normality and pathology, argued that pathology can only be created alongside such a norm, as people identify it and decide that they prefer it to the apparent alternatives. Creating a pathology is thus a process of assigning a negative value to situations that diverge from those that are preferred. Canguilhem was explicit about such negative valuations; he described normalization as a polemical process, arguing that those who prefer the norm view divergence from it not with indifference but with repulsion and hostility. According to Canguilhem, the norm aligns with what is right, or what is straight, and is opposed by what is pathological, or what is "twisted, crooked, or awkward."[6]

Ascher created templates illustrating the twisted, crooked, and awkward character of inefficient human motion. He used what his colleague Helmut von Bracken called a cinematographic method, similar to the well-known serial photographs of Étienne-Jules Marey and Eadweard Muybridge, in which he fitted workers with lighted points whose paths were captured in split-second intervals. He was then able directly to compare how highly productive and less productive workers moved. Skilled and efficient workers followed smoother, straighter, or shorter paths of motion than did unskilled or inefficient workers, and workers using supports exhibited less shaking and swaying.[7]

Ascher illustrated his technique in a motion template he made of apprentices training to become smiths. It shows the paths of their wrists, elbows, shoulders, hips, and knees during a hammer strike and, when read from left to right, shows how practice and training could overcome crooked and awkward movement. An apprentice's untrained wrist followed a long path with several sharp curves, at two points doubling back on itself. The apprentice's untrained elbow followed a more ranging route with even sharper corners: the path looks almost like a triangle. His shoulder, hips, and knees followed no discernible path at all. In contrast, the wrist, elbow, and shoulder of the skilled smith traced smoother, gentler, and straighter paths; there were neither the sharp angles nor the doubling back that character-ized the beginner's strike. The jumble of points that marked hip and knee action were replaced by a single point, indicating that the hips and knees remained still, a sign that the worker was practicing a true handicraft. Similar templates of filers and joiners illustrated the more direct motion of skilled and efficient workers. Less-skilled workers' arms, hands, legs, and torsos ranged widely and indetermi-nately; more practiced workers demonstrated greater control.[8]

These results concerned Ascher, especially because they occurred at the inter-section of human habits and new industrial technologies. For him the new me-chanical age raised the possibility of improved human efficiencies and a longer effective work life for many people, and he believed a full understanding of the differences between people and machines would make it possible to exploit the newer machines fully without subjecting workers and their bodies to additional and unnecessary stress. Ascher had been impressed by augmented efficiencies among weavers he studied using automatic looms, which he compared with Max Weber's findings in a study of weavers using traditional technology in Mönchen-Gladbach and with Adolphe Quetelet's more general observations about the rela-tionship between a worker's wages, age, and strength. Quetelet had observed that at about age thirty-five a worker's lifting ability and wages both declined precipi-tously, whether or not the work demanded heavy lifting. Weber's study of tradi-tional weavers corroborated this finding: at about age thirty-three their earnings, too, went into a steep decline. In contrast, Ascher found that weavers using auto-matic looms experienced a small but steady increase in wages from age twenty through at least sixty.[9]

Ascher's results paralleled an earlier development in industrial technology: an emphasis in machine design and theory that identified the most efficient mech-anisms as those whose motion was most controlled. One of John Smeaton's most important contributions, in his celebrated eighteenth-century experiments on waterwheel efficiency, was a method of preventing turbulence by controlling the motion of the water. Gérard-Joseph Christian, director of the Conservatoire des Arts et Métiers during the Bourbon Restoration, had described the most per-fected machines as those whose nonworking parts remained rigorously immo-bile. The importance of motion for productive machines was affirmed by Franz Reuleaux, professor of theoretical mechanics and director of the Industrial Insti-tute of Berlin, whose kinematic theory of machines helped transform the study of mechanisms in the last decades of the nineteenth century and remains in-fluential today. Reuleaux distinguished between determined and unconstrained

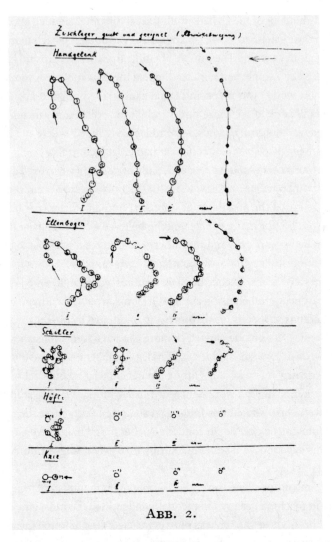

ABB. 2.

Ludwig Ascher's motion curves of smiths making a hammer strike, showing paths of motion for wrist (*top*), elbow, shoulder, hip, and knees. The smoother and straighter paths on the right represent the motions of more skilled smiths; note also that the hips and knees of the more skilled workers remain nearly stationary, a sign that they are practicing a true handicraft. Ludwig Ascher, "Der Einfluss technischer Verbesserungen auf die Gesundheit des Menschen, insbesondere des Arbeiters," in *Resoconti del Congresso*, vol. 2: *IIIo Congresso del Internazionale di Organizzazione Scientifica del Lavoro* (Rome, 1927), 2:565. Photo by James Alexander.

motions, or between control and turbulence. A machine performs work only through "determinant motions." It uses, and indeed compels, forces of motion to perform mechanical work, but such motion must be completely determined or constrained by a combination of mechanical and resistant structures to prevent turbulence, or what Reuleaux called "disturbing motion." Reuleaux's requirement that all motion be controlled was firm: he described the moving part in one class of kinematic pairs as "imprisoned" and prevented from "every motion except a single one."[10]

Fatigue was the important concern with human motion, and as in machines, it had to do both with wear-and-tear and the waste of motive power. As Ascher saw it, one of the greatest limits to human efficiency was fatigue, which was embodied, for him, in unnecessary motion. Even a skilled or naturally gifted worker's movements could be deranged and made inefficient by fatigue. The problem had long occupied labor researchers, many of whom described it as an obstacle to progress because it limited people's ability to tap into the vast reservoirs of energy stored within their own bodies.[11] Ascher's solution to the problem was twofold and turned on a distinction between parts of the body that were supposed to move in productive work and those that were supposed to remain at rest. For parts that were supposed to move, Ascher prescribed training to improve skill. For parts of the body that were to remain at rest, he prescribed supports to help bear their weight and prevent trembling and stress. Allowing only the working parts of the body to move, while supporting its stationary parts, would combat fatigue and thereby increase productivity and efficiency.[12]

Keeping still thus became a crucial component of human efficiency. What did not move would neither cause fatigue nor waste energy: this was Hermann Hebestreit's crucial contribution. In 1930 Hebestreit was fresh from his clinical year at Frankfurt University's Institute for Animal Physiology, where he had written a dissertation on how bodies recovered from physical labor;[13] he published a lengthy article the same year on the physiological principles behind worker posture, which coincided with the reorganization of the worker seating exhibit in its traveling form, and described the research upon which it was based. Hebestreit had designed a research program to investigate the physiological principles behind Ascher's templates, undertaking a series of studies to compare the moving parts of the body with those that were to remain at rest. Like Ascher, he had used serial photography to analyze motion and energy expenditure, supplemented by studies of skeletal structure and analyses of workers' centers of gravity and their use of supports. His diagrams illustrated the relationship between posture and forces on the body, forces that he described as pathways siphoning off motion and thus energy. Wasted motion caused fatigue; that it also wasted energy was made explicit in Hebestreit's research.[14]

Hebestreit was especially concerned with static work, in which portions of the body, such as the torso or legs, remained immobile. Static work also drained energy and led to more wasted motion as fatigue set in and workers swayed, changed position, or found their muscles trembling. Hebestreit noted that posture was particularly important in preventing inefficiency and fatigue among seated workers. When a worker had to stand, it was best if the center of gravity of the torso could be located directly over the hip joints, or over the fulcrum or pivot of motion. When a worker sat, that fulcrum was somewhere else: in the tuberosity of the ischium, or in the lower pelvis in what physical therapists and dancers sometimes called the "sitz bones." The efficiency of static work improved, and motion and energy loss were reduced, when the torso's center of gravity lined up directly over the fulcrum, or as close to the vertical position as possible. Chairs with lumbar supports helped hold the torso in the right place; because it was a heavy part of the body, leaning on the arms or slouching to hold it steady used large muscles and wasted motion and energy.[15]

Hebestreit's work thus supported a general conclusion about which parts of a worker's body should move: the smallest ones. Bigger body parts were heavier, with more massive muscles; they made longer arcs of motion and used more energy and were best supported and held at rest. Efficient bodies did this naturally anyway, Ascher argued.[16] Ascher had compared two seamstresses: the least productive continually leaned forward to pick up new pieces of fabric, while the other kept the pile close at hand and her torso still. Barbara Orland has uncovered similar comparisons in her study of rationalization and women's work, in photographs depicting it as inefficient for a seamstress at a machine to lean forward and rest the mass of her torso on her forearms. In contrast, a woman of efficient posture used the fine muscles of her hands to inch the fabric along, her torso erect and supported.[17] Her motion was more controlled and covered a smaller range.

A Crisis of Disorder: Rationalization, the RKW, and the Need to Manage

Ascher's and Hebestreit's work contributed to the ambitious rationalization movement in German industry in the 1920s. Rationalization, to many, offered an efficiency solution to economic and social turbulence in a country recovering from military defeat and revolution at home and facing international condemnation as a war aggressor and a potentially catastrophic reparations bill. What form the efficiency solution should take was the subject of ongoing controversy. The rationalization movement continued to gain momentum throughout the decade, but it was never united and defied easy description; groups with widely differing ideas on the causes and cures of economic woes, and on the social role of economic affairs, had widely different visions of rationalization. What coherence the movement did have lay in the shared belief that rationalization "was a panacea for the ills besetting the German economy" and the key to improving the economic and social order.[18] Rationalization, and its main institution, the Reichskuratorium für Wirtschaftlichkeit, illustrates a dynamic efficiency used chiefly by industrial managers in an attempt not only to manage but to transform a turbulent society.

The crisis was real enough: Weimar Germany's was a "sick economy."[19] Astronomical inflation early in the decade was followed by spikes of unemployment higher than any that had ever been seen before, accompanied by only limited and fitful industrial growth. The pattern was of crisis and stagnation: war, revolution, and the founding of the republic itself; hyperinflation in 1922 and 1923; brief recovery after currency stabilization in 1924; staggering levels of unemployment in 1926; a brief period of productivity growth accompanied by high unemployment from 1927 to 1928; and then the final crisis of 1929–33, with the slide into world depression, accompanied by unemployment of nearly 30 percent and chaos in the streets, in deadly battles between political parties and police.

Rationalization was to have prevented all this. The term itself, *Rationalisierung,* had first achieved wide currency as a description of rationing production in Germany's heavy industries following the extreme inflation of the early 1920s, and Carl Friedrich von Siemens gave a widely quoted definition in a 1921 speech: "Rationalization," he argued, "is the adoption and employment of all the means of increasing efficiency which are furnished by technical science and systematic organization. Its aim is to raise the general level of prosperity by cheaper, more plentiful and better quality goods."[20] Siemens's flat statement belied rationalization's larger scope: its goal was transformative and ambitious and was connected to broader plans for social engineering. In Siemens's own business, the Siemens's electrical concerns, programs to rationalize production were accompanied by new social policies that structured employees' leisure time and family life, including company-organized housing and vacations. Herbert Hinnenthal, RKW business manager, defined *rationalization* as "everything that could restore equilibrium," and Robert Brady, the American business analyst whose contemporary studies remain indispensable, described rationalization as seeking the "'efficiency' key to orderly social and individual life," economic stability almost invariably given as its goal.[21]

Crucial to rationalization was a concept of flow. It could describe the assembly line and other practices for keeping the productive works in continual motion, and many in Germany were impressed to learn of such methods at use in the United States in the prosperous Ford Motor Works. But *flow* also carried another meaning, referring not to specific techniques but to a more general ideology of undisturbed production. If the solution to social and economic crisis lay in the raising of living standards through cheaper and more plentiful goods, then whatever imperiled production further imperiled a society already in crisis. Many technical measures were undertaken to streamline production, including standardization in many forms, of work schedules, parts and sizes, and methods of production; widespread adoption of new cost-accounting methods; and a host of technical measures to reduce waste: coordination of transportation and delivery systems, specialized technical education, and the introduction of specialized machinery, including new methods and equipment for the winding of motors at Siemens and pneumatic picks for winning coal at the face in the mines of the Ruhr.[22]

Among the most intractable impediments to smoothly flowing production were people themselves, and rationalization's personal and social effects and implications made it bitterly controversial. The social and personal implications of technical rationalization were indirect; rationalization required new patterns of often more intense work and made many jobs redundant, but it was not aimed primarily at workers themselves. Social rationalization was. Its goal was to reduce human disturbances to the flow of work by engendering cooperation and loyalty and by preventing personal life and personal habits from influencing production. Carola Sachse has described how the concept of flow was integral to social rationalization and efforts to minimize disturbances caused by human factors at Siemens, and Heidrun Homburg has characterized worries about human disturbances as part of an ideological position that production should be kept free of worker friction or resistance.[23]

The rationalization issue thus resonated far beyond the technical realm and raised fundamental questions about the relationship between industry and society. At issue was whose vision would direct the remaking of the German economy, and such high stakes made rationalization bitterly controversial. Members of the Social Democratic Party saw in it a way to increase productivity without eliminating the eight-hour workday, and the socialist Rudolf Hilferding celebrated it as a step toward a planned, cooperative economy, describing institutions such as the Reichskuratorium für Wirtschaftlichkeit, the leading institution of rationalization, as evidence of an "interpenetration of economy and state" and proof that "the running of his business is no longer the private affair of the entrepreneur."[24] Representatives of labor denounced the movement as intensified capitalist exploitation, and many workers objected to particular features of the movement, such as the use of stopwatches in time and motion studies, the speeding up of work, and an increasingly minute division of labor accompanied by falling piece rates. Even groups who agreed in principle on the desirability of rationalization did not agree on the particular form it should take.[25]

Efficiency measures largely remained tools of industry.[26] Some industrial representatives found in rationalization "a new weapon against the finance and welfare policies of the Weimar Republic," policies reflecting the welfare state created by the Weimar constitution and extending the ethos of social insurance laws first passed in 1881, in which the Wilhelmine state had recognized that neither private charity nor laissez-faire development could guarantee adequate living standards or opportunities for its citizens. Rationalization offered industry a way to describe its vision of a society revived and made newly prosperous but without governmental interference. Debates over rationalization were especially rancorous after the deepening financial crisis of 1929, when the concept became a rhetorical tool of both employers and organized labor in battles over political reform.[27]

What coordination was possible in such a charged atmosphere was provided by the Reichskuratorium für Wirtschaftlichkeit. The RKW stood at the center of a comprehensive web of institutions devoted to rationalization that developed in Germany following the war and which by the mid-1920s had become widely known as a clearinghouse for information and recommendations on all aspects of rationalization. Its prominence rested on a key strategic decision made at its founding in 1921, when it was chartered by the reich but provided with only minimal funds: that it would remain an advisory body, without rule-making or regulatory authority. Many in industry supported it only with the assurance that it would not take on governmental functions. The RKW's original charge was to extend standardization measures adopted in World War I, and it was little known until 1925, when its directors, Carl Friedrich von Siemens and Carl Köttgen, secured more generous funding from the government and reorganized it into a prestigious partnership between industry and the state and with a much more ambitious mandate: to help remake the German economy through rational management and industrial cooperation and, by so doing, help ameliorate the inequalities and the possibility of social strife that worried many of Germany's policy makers and social observers. Its members included other organizations such as the Committee for Economical Production and the Committee for Economical Management, and it worked closely with the Reichsverband der Deutschen Industrie (National Federation of German Industries). It became the central co-ordinating body for rationalization programs across the reich, promoting standardization, studying technical problems in various industries, disseminating information on organization and management issues, and facilitating the work of other bodies devoted to specialized aspects of rationalization. It sponsored conferences, published and distributed literature, and undertook research into technical issues including methods of packing and shipping goods, standardized drive systems, and cost-accounting and budgeting practices. The RKW's goal, as described in its numerous publications and reiterated in speeches by its leaders, was dynamic and transformative: to stimulate prosperity and do away with need.[28]

The RKW formed a nexus between government and industry. Its ties to the Siemens's electrical concerns were particularly close; indeed, the House of Siemens and the RKW shared the same leadership for more than a decade. Carl Friedrich von Siemens, Siemens's CEO, served as chairman of the RKW for its first ten years and was succeeded in 1931 by Carl Köttgen, who was also general director of Siemens-Schuckertwerke, Siemens's heavy-current division. Köttgen had in fact been acting as the RKW's director since its inception, and by the time he stepped down in 1934, the institution had spent thirteen years under Siemens's leadership.[29]

The RKW never enjoyed a stable period of normal operations. Shortly after its reorganization in 1925, it was implicated in a "rationalization crisis": a period of high unemployment for which many blamed rationalization, arguing that efficiency measures had thrown people out of work by replacing them with machines or making their positions redundant.[30] Thirty to 34 percent of unemployment in Ruhr coal mining in 1926 and 1927 has been attributed to rationalization, and unemployment through de-skilling and redundancy in the electro-technical industry had by 1927 resulted in a marked restructuring of the labor force. Unemployment remained extraordinarily high throughout the rest of the Weimar period, exacerbated by the crash of the U.S. stock market in the fall of 1929 and the withdrawal of the many American loans that had helped to sustain the Weimar economy. The rate of unemployment rose from 8.5 percent in 1929 to nearly 30 percent in 1932, when more than six million people found themselves officially out of work. Unregistered and part-time workers may have numbered between one and three million who were also unemployed.[31]

The experience of the House of Siemens and the electrical industry illustrates the parameters of the crisis. The heavy-current sector was hit first and hardest, as municipalities experienced fiscal crises themselves and were unable to afford purchases for infrastructure development, such as new steam turbines for power plants. Municipalities that had floated bonds on the U.S. market to finance such purchases faced bankruptcy as payments ceased, even as their tax income fell and the demand for social service spending grew. In 1929 Siemens was performing at 78 percent of capacity, by 1932 at only 42 percent. By the summer of 1930 Siemens-Schuckertwerke had laid off 32 percent of its manual employees, and between November 1930 and December 1931 Siemens cut wages nearly 30 percent. By 1933 unemployment in the electrical industry had reached 38 percent, and many people in the economy at large had been without work for five years and subjected to the increasingly degrading requirements for unemployment or welfare subsidies or for assistance from private charities.[32]

The magnitude of the crisis is difficult to overstate, its social and political dimensions overwhelming. Peter Stachura describes the depression in Germany as "an unprecedented catastrophe from virtually every standpoint." Mass unemployment was the critical problem, and it presaged the crisis that brought about the collapse of the republic. The entire cabinet of Chancellor Hermann Müller resigned in March 1930 over its inability to agree on financing for unemployment insurance, and Heinrich Brüning, Müller's successor, removed policy making from the parliamentary arena in an effort to circumvent continued impasses, in effect bringing Germany's republican experiment to an end.[33]

Rationalization seemed to have been truly discredited by this last crisis, being either a complete failure or entirely beside the point. The RKW itself faced public condemnation, as critics charged that rationalization had helped to cause the depression by putting thousands of people out of work. Laborers were increasingly hostile to rationalization, and even those in industry who remained committed to it in principle blamed the German movement for having failed to provide the necessary social planning.[34] The Reichstag cut the RKW's budget drastically for the 1929–30 budget year, authorizing only 500,000 reichsmarks, when in 1927–28 it had authorized 1.45 million; only 240,000 reichsmarks were authorized for 1930–31, and only half that, 120,000 reichsmarks, for 1931–32. Even reduced funds were made contingent on the RKW turning its attention away from the primarily technical and economic measures that had occupied it to questions important to workers, including the effect of rationalization on social and worker health.[35] The RKW fielded more modest projects in 1930 and 1931, principally publications, public lectures and discussions, and exhibits such as the one on efficient worker seating. Publications continued as planned: the serial *RKW-*

Nachrichten; brochures on business statistics, on rationalization and housework, and on the human aspects of rationalization; and a third edition of the exhaustive *Handbook of Rationalization* under preparation. Ascher led one of five public discussions, his on the practical principles of work posture; the five events each drew as many as eight hundred people. The RKW also sent delegates to Geneva to a budgeting conference and opened a line of research on the trades.[36]

The most visible response at the RKW was a new highlighting of its programs on human factors, as it redirected its dynamic hopes for social transformation to workers themselves. It now claimed, as in its 1930 annual report, that making the best use of human labor power was the first and foremost principle of rationalization, and in public forums its members began to argue that rationalization was indeed grounded in human work.[37] The head of the RKW Committee on Economical Management called this emphasis on human factors "the newest slogan in a movement already packed with jargon," but it did indicate a shift in emphasis toward areas in which rationalization might show itself immediately relevant.[38] An RKW-sponsored exhibit in Düsseldorf in 1930 explicitly addressed the question "of what use is rationalization when [unemployment] has thrown thousands into the streets?"[39] Unemployment was an undisputed by-product of rationalization, the catalog admitted, arguing, however, that it was temporary and would be overcome by adequate social rationalization, beginning with guidance in job choice, job training and placement, and integration of the working person into the economy, in effect managing an employee from school days through retirement. Rationalization's transformative and dynamic goals became particularly apparent in discussions of workers; Wilhelm Eggert, of the Executive Council of the General Confederation of German Trade Unions, was quoted in a plaque prominently displayed in the center of the Düsseldorf exhibit, telling the working classes that they were "compelled to take notice" of rationalization, for it "reaches deep into the relations between people and work, citizens and state, and local and world-wide economy." These relations were to be managed as part of a larger effort aimed at "engineering the new worker," as Mary Nolan has put it.[40]

The climate, from 1929 onward, was one of unrelenting crisis. Those who were unemployed took to spending full days in the streets and parks, their living quarters often too small and crowded to accommodate them during the day and their funds too limited for them to pass the time sociably in a tavern or café. Young people also took to sleeping out. Those receiving assistance, whether from the state or from private charities, experienced the increasing humiliation of home inspections, arbitrary check-in requirements, and full days spent waiting in lines hundreds of people long, sometimes in halls and vestibules so crowded that people fainted from heat and lack of oxygen.

One of the most disturbing features of the disaster was violence in the streets, organized by the paramilitary wings of political parties. The violence was not only political but often despairing, especially on the part of militants in the Kommunistische Partei Deutschlands (KPD, Communist Party of Germany), who found themselves defending their neighborhoods against both the police and the Sturmabteilung (SA), the militia group of the Nationalsozialistische Deutsche Arbeiterpartei (National Socialist German Workers' Party [the Nazi SA]). Violence marked the opening of the Reichstag in October 1930, when the National Socialists for the first time gained a position of real power, having become the parliament's second strongest party; there were mass demonstrations in Berlin, and Nazi youths smashed the windows of Jewish-named department stores.[41] Harry Kessler's celebrated diaries charted a fearful chronology: gangs of homeless children wandering, thieving, and looting in the working-class sections of Berlin; political parties training their own troops for threatened civil war; political street fighting in the summer of 1932, claiming seventeen lives in the poor working-class district of Hamburg-Altona on July 17 and, over the course of the season, a total of nearly a hundred lives, with more than a thousand injured.[42]

Unemployment contributed to the instability. Many in industry saw unemployment as the greatest threat to social order and believed that workers, without the discipline and monitoring provided by a job, would become degenerate and lazy and tempted to take up political radicalism.[43] The experience of the Communist Party partly bears this out, with respect to radicalism but not to degeneracy or laziness: as conditions worsened, the KPD continued to gain votes, and nearly 60 percent of its street fighters were unemployed. But those fighting were not *Lumpenproletarier*. The percentage of unemployed among the street fighters was high relative to unemployment in the labor force generally, but it bore a close parallel to the structure of unemployment among those in the skilled trades, and an analysis of occupational data given to police by Communists arrested for street fighting in Berlin suggests that in previous generations the fighters would have been "members of the stable middle to upper ranks of the working class."[44] Unemployment had radicalized not the rabble but skilled members of the trades who had formerly commanded public respect. That they were significantly represented among those engaged in street violence underscores the magnitude of the social disintegration in the last years of the republic.

The RKW represented a movement whose great ambition to remake economy and society had failed, and failed terribly. The efficiency solution had not managed to produce more goods more cheaply, it had not relieved poverty and want, and it had brought neither general social stability nor harmony and cooperation between workers and their employers. By late 1932 the situation had become intolerable, violence undermined any possibility of fair elections, and the streets of Germany's capital city had become unsafe.

But the situation had already deteriorated by early in 1929. Although the crash of the U.S. stock market exacerbated the crisis, it did not cause it; unemployment was already high, and there had already been significant violence. May 1929 in fact came to be known as "Bloody May," when fighting between police and Communists claimed thirty-one lives in the impoverished district of Wedding, in the northeast quadrant of Berlin; hundreds more were wounded, and more than a thousand were arrested. It was in this month, quietly and across town, that the RKW opened its exhibit on efficient worker seating.[45]

The "Work Seats and Work Tables" Exhibit
Managing Disorder

The exhibit "Work Seats and Work Tables" opened in Berlin in May 1929 at the Labor Protection Museum on Fraunhoferstrasse in the affluent residential and shopping district of Charlottenburg. For a modest entry fee of thirty pfennig, visitors could study full-scale installations of efficient seating and working arrangements currently in use and details of the medical research on which they were based. The exhibit was so successful that a traveling version was unveiled at the Berlin Welfare Ministry in May 1931 and scheduled to stop in Karlsruhe, Kaiserslautern, Nuremberg, and Leipzig. The traveling exhibit met with favorable reviews in the labor, engineering, and medical presses, and its sponsors solicited invitations from other prospective hosts.[46]

The seating arrangements on display in the exhibit were designed to inhibit: the point, as the reviewer for a prominent journal of labor health put it, was that "all motion not directly necessary for work should be eliminated." Specially arranged work spaces, incorporating arm- or footrests and including properly configured stools, seats, and tables, would eliminate unnecessary movement.[47]

The traveling exhibit was divided into two sections. The first showed the medical and physiological foundations of worker posture and included illustrations of women stooping beneath heavy burdens, life-sized portraits of seamstresses showing fatigue and stress, and diagrams of seated skeletons illustrating the appropriate points of the body for bearing stress or force. The second section demonstrated current practice. It showed how worker seating had evolved from old work stools to modern chairs, with fixed or adjustable back and arm supports and able to swivel or roll; it also reminded visitors that efficiency required careful attention to other things as well: the supply and removal of materials, adequate lighting, and protection from accident. Model designs from the electro-technical and cigarette industries illustrated how worker posture apparatus might be deployed in factories, and examples from the Reichspost, the postal service, showed how they might be used for office or other nonfactory work. Siemens contributed models and photographs, including an installation of work seats developed at Elmowerk. Adjustable armrests could be added, and the seat could be mounted on a fixed bar, allowing it several feet of lateral movement.[48] The supports and constraints on view were designed to guide workers' bodies along efficient and predetermined paths.

The choreography of motion control took two forms: mechanical devices to inhibit unproductive motion, such as supports and backrests, and training and education designed to yield efficient motion through workers' own self-discipline and practice, or drill. Among the RKW's educational materials appeared a plaque entitled "The Ten Commandments of Work Seating," which outlined an elaborate choreography conscripting the entire body, through directions governing posture and seating from head to foot. The spine was to be supported with the trunk erect and arms left free; the upper legs were to be supported from the hip to just behind the knee so that the worker need not push against the floor to stay seated; the seat must be high enough to keep the knees at approximately a 90° angle; the shoulders should not slump forward; the head should not hang down over the task; feet should be placed flat, on the floor, footstool, or pedal; and the arms should be supported if the work was principally done by the hands. Arms were not to be extended far in front of the body, and upper arms were to be nearly parallel to the torso.[49] This body, by and large, was not supposed to move.

The choreography emphasized remaining still, or at rest. Things that were not moving wasted no energy. Leaning forward and reaching was labeled *Falsch* (false) in one plaque; also *Falsch* was a woman leaning back in her chair. *Richtig* (correct) described a woman sitting up straight and holding her upper arms parallel to her trunk, her seat and back following the contours of a chair similar to the Elmo work stool. Choreographed routines helped workers endure the stresses of stationary labor by allowing them to move in ways that maintained their ability to work. Workers were directed to use their muscles as pumps, periodically raising and

"The Ten Commandments of Work Seating." *RKW Nachrichten* 5 (1931): 124. Courtesy of the Staatsbibliothek, Berlin.

lowering their arms to keep their blood flowing freely, and cautioned against leaving their arms still or hanging and against holding things aloft for too long. They were also told to stretch out their feet occasionally.[50]

The exhibit and educational materials were not merely suggestions; many of the principles and structures on display were already in use in cigarette-making factories and in the offices of the postal service. Siemens's engineer C. W. Drescher photographed worker posture measures also in use in Elmowerk, where managers had adopted them earlier in the 1920s.[51] Drescher's photographs show different combinations of motion-inhibiting mechanical structures intended to augment efficiency by containing workers' movements in a confined and limited space.

In one photograph a worker sat at a grinding machine in a work stool with an adjustable back and of adjustable height, designed for use at tables of different heights and by people of different sizes. Only the worker's forearms and hands were engaged in immediately productive work, and they were left free. Her feet rested firmly on a platform beneath the table holding the machine, her knees at a 90° angle. Another worker, at a spindle boring machine, also sat on an Elmo work stool, this one mounted on a rollbank. The height of the seat was again adjustable, as was the spring-action back support, and the rollbank had protective edging on three sides, to prevent clothing from snagging the rollers. Forearms, hands, and upper arms were expected to engage in immediately productive work,

Worker in a specially designed support, at an inclined press, in the Siemens's Elmowerk. C. W. Drescher, "Arbeitssitz und Arbeitsplatz: Entwicklungsbeispiele aus der Praxis der Siemensbetriebe," *Arbeitsschutz*, June 15, 1929, 170. Courtesy of the Staatsbibliothek, Berlin.

and they were left free, as were the feet, for rolling. A worker at an inclining press used a specialized work stool. The seat was also fully adjustable, and because the operator's required posture was unusual, being midway between sitting and standing, it provided additional supports for thighs, hips, back, and feet to help keep the worker from sliding downward. This posture was accommodated by the machine itself, which had been designed to allow the worker to operate it from above, outside what Drescher called the "accident zone." The operator used hands and forearms, with some upper arm motion; all other body parts were to remain at rest.[52]

Efficiency and Mechanical Discipline

The worker seating exhibit illustrates a managerial approach to individual efficiency, designed to promote stability among workers as part of a larger effort to transform a society in deep distress. It portrayed a view of efficiency that relied heavily on managerial control, in the structural supports themselves and in the motion templates upon which they were based. And while it is tempting to argue that such control may be balanced against the potential health benefits of such measures and against the good intentions of those who introduced them, it remains true both that the supposed health benefits have not withstood the test of time and that those who introduced such measures routinely denied doing so for altruistic reasons. They defended them as sound business practice.

The traveling exhibit displayed a therapy while also illustrating the developing norm of efficient human motion. It suggested training and practice to increase a worker's efficiency through increased skill and offered seats and workspaces designed to help eliminate "all motion not directly necessary for work."[53] The therapy also appeared in educational materials and in reports on the implementation of worker posture apparatus in the Elmowerk.

To speak of therapy underscores the disciplinary function of the exhibit and the apparatus it displayed. In deploying a therapy, researchers and managers demonstrated their belief that workers posed a problem that could not be solved without outside intervention and that the problem was not merely personal, for it had larger social effects. Workers, through inefficiency, wasted valuable energy that could not be recovered and hindered efforts to help ameliorate through productivity the acute economic and social difficulties of the late Weimar years. The techniques and motivations behind the exhibit reflect a therapeutic orientation: an attempt to steer situations in different directions than they were already headed, through instruction or training or through material interventions, described as benefiting both the recipient and the society of which he or she was a part.

Behind the exhibit lies an ideology of control. It was a microcosm of mechanical predictability in an increasingly chaotic time. The exhibit opened just as the RKW began to face the failure of its larger goals and illustrates what James C. Scott has called "miniaturization": "the creation of a more easily controlled micro-order" by social planners and engineers who find their larger aims thwarted. By 1929 the RKW had failed to achieve its broader goals: rational management and industrial cooperation had not transformed the German economy, and depression was deepening and unemployment rising.

The very techniques behind the exhibit illustrate control. Ascher's templates identifying the source of fatigue as unnecessary movement were both analytical tools and practical models to guide the design of more efficient apparatuses. They aided not only in analyzing motion but also in controlling it. Scholars have recognized the power of representing a complex and moving, or changing, system in stasis; it is a technique for gaining control. Representing processes in two dimensions and on paper was an essential step in gaining control over production processes, as was measurement itself. "Technological representations," Steven Lubar argues, "make it easier to bring technological actions under the control of authority."[54]

The structures and supports on display in the exhibit illustrate mechanical discipline: the use of mechanical means to ensure that workers followed the rules of efficient movement.[55] They were designed to create and enforce a balanced and mechanical form of efficiency: regular motion, even and unperturbed by fatigue. In 1933 the analyst Robert Brady described how rationalization had rendered the German worker passive and susceptible to discipline, he or she having become "a functioning datum fitted, much as any other piece of machinery, into a mechanically controlled and regulated routine."[56] Brady also believed that the historical meaning of rationalization would be found in the confluence of capitalist economic organization with fundamental tendencies in machine design, particularly in a machine's standardizing function and its promise of regularity and uniformity.[57] Such regularity and uniformity were increasingly achieved through mechanical structures that controlled how a machine's working parts moved.

The mechanical discipline of the structures was not a matter of educating workers about rules of movement. Training and practice, following rules, and maintaining an open and erect posture are readily recognized as aspects of an internalized discipline; they describe, for example, the practice of military drill.[58] The disciplinary function of the mechanical structures is less immediately apparent and requires engaging how external physical structures influenced people's personal movements. The devices on exhibit were designed to inhibit movements that wasted energy and interfered with productivity—movements caused principally by fatigue, which training and rules could not prevent. Mechanical discipline here extended the metaphor of the human machine into actual practice, in the form of structures that deployed the body in mechanical ways.

By inhibiting even their unintentional movements, the mechanical discipline of the efficiency apparatus underscored the lack of autonomy of the workers within it. Individualized movements were outside the choreography and against the apparatus, and they were proscribed. As long as workers were within the apparatus, they were expected not to indulge in autonomous gestures of the sort that displayed their individual qualities or character. Waving to a friend or stretching, bringing one's arms high above: these were disturbing motions that violated the rules of choreography, but they were not prohibited by the apparatus itself. What the apparatus did make difficult was deviating from its own dictates: wobbling despite backrests or trembling despite arm supports. The apparatus helped make clear that deviations—whether slumping, leaning, crossing the legs, or swinging the arms—were likely to be intentional. As described earlier, such actions were labeled *Falsch*. Intentional but unplanned and nonchoreographed movements were disturbing and not immediately necessary for the work at hand: they were pathological.

The desire for control inherent in these efforts may be balanced against measurable gains in productivity and against potential health benefits, although both are difficult to assess. Whether the measures did enhance productivity remains unclear: Wilfried Feldenkirchen has tried to determine the effect of these and other rationalization measures on Siemens's productivity but finds it impossible to link indices of performance to particular technical or organizational measures.[59] Evaluating how committed motion researchers were to improving worker health and safety is also problematic. Many were genuinely concerned. An example may be found in the photo of the woman in the incline press presented earlier; the design of her seat and of the machine suggests a caring orientation that recognized both her need for comfort and how vulnerable to accident she was. The press had protection built into it. Ascher, in particular, believed that, by uncovering principles already in use by healthy people who were naturally efficient, graceful, or skilled, his work had far-reaching implications for health and wellbeing. Hebestreit, too, described efficiency as a key to promoting health, but he meant something different than Ascher did; Hebestreit was referring to an increase in "manageable labor power."[60] In clear contrast to Ascher, the chair of the RKW's economic management committee explicitly separated philanthropic or humanitarian concerns from productive ones: managing human factors and overcoming "human friction" were not acts of altruism, he argued; they were merely a necessary part of rationalization.[61] Werner von Siemens himself had insisted early on that "healthy self-interest" and not altruism underlay his company's extension of health and welfare benefits to employees, an orientation that continued into the period of rationalization.[62]

Health alone, then, need not govern how one analyzes the posture measures. There is now general agreement that good posture enhances health, but not all of the recommendations of the exhibit and related materials have stood the test of time. Many repetitive stress injuries are now described as being due to repetitive small muscle movements, or to precisely the sorts of motions advocated by the researchers discussed here. Even if the worker posture measures were beneficial, that they may not have been designed primarily for comfort or for health suggests that such benefits may have been secondary and legitimizes analyzing the measures in other terms. It justifies examining them in the terms of machine design, the terms that Brady suggested.

The efficiency structures, when distinguished from training, practice, and rules, most clearly illustrate how mechanical discipline limited worker autonomy. Participating in training and practice and following rules may be seen as acts of human will, people bringing their behavior into conformity with an educating or regulating authority, whether freely or, in a precarious labor market and in fear of losing their jobs, grudgingly or in despair. In contrast, the posture apparatus, once a worker had entered it, constrained the worker unless he or she made an active attempt to resist it. Mechanical discipline relied on such structures; it was external to the worker and resided in the apparatus, was engaged when the worker got in, and operated as long as he or she stayed there.

The emphasis on efficient worker posture was not uniquely German, but the German context underscores a more general sense in which the discipline of efficiency, with its emphasis on industrial regularity and uniformity, may be seen as a method of overcoming workers' personal autonomy and individuality. This assessment is particularly stark in the work of some scholars of Weimar Germany and the rationalization movement who have argued that workers found their personal autonomy compromised as opportunities for action and resistance were eroded. Mary Nolan concluded that the working classes in general, along with the German labor movement, "were the principal losers in the bitter battles surrounding rationalization" and that resistance became "extremely costly and unlikely" because it threatened job loss in a period of high unemployment. She also noted that, although the movement did not displace skilled workers, it can be said that work for both skilled and unskilled workers intensified and became more oppressive for many of them. Peter Hinrichs, in a study of the development of industrial psychology, concludes that industry and business claimed to be "humanizing" work primarily to justify measures that enhanced their own authority. Heidrun Homburg has perhaps the harshest assessment, concluding that Siemens, Köttgen, and the RKW took an ideological position that classified workers as objects and considered the "concrete individual person as sand in the . . . machinery of rationalization."[63]

Similar arguments may be made beyond the German context. Andrew Pickering writes more generally of industrialization as an "attack on the body—a devaluation of human agency in relation to the agency of the new machines," and James C. Scott argues that narrow and planned environments "foster a less skilled, less innovative, less resourceful population," which, paradoxically, ends up creating workers who are more likely to require the kind of close supervision that planners assumed they needed at the start.[64] The idea of mechanical discipline, which turns on the extension of the metaphor of the human machine from an intellectual conception into actual working practice, and which in this case draws an analogy between motion control in machines and motion control in people, helps illustrate how efficiency measures treated workers as deeply passive and tried to enforce that passivity by controlling how they moved.

In the worker seating exhibit efficiency again had two meanings. It described the work of those whose motions were directed along the narrowest paths, indeed, those who moved the least. It also described the result and goal of a mature economy and the aim of the industrial managers and its sponsoring institution, the RKW: a healthy and stable distribution of the necessities of life. As the RKW failed in its larger aims, it continued with efforts to manage the individual bodies of workers, no longer claiming it could transform their lives or transform society but simply to get them to sit still.

The developing concept of efficiency was evident in concerns about worker posture and motion in late Weimar Germany. An industrial norm of efficiency, which required straight lines and smooth curves of motion in machines, was paralleled in a pathology of human inefficiency. The solution was a model of rationality, planning, and control: people made efficient through the discipline of training, practice, and following rules and through a mechanical form of discipline that relied on structures and supports to constrain how they moved.

Behind the worker seating exhibit lay the question of discipline. The disciplinary function of training and drill and rules is readily apparent; the mechanical discipline of the seats and supports is less easily discerned but can be seen when they are analyzed in the mechanical terms of motion control. Together, the supports and structures, along with the choreography of motion and rest, were designed to make workers efficient by eliminating unplanned, and thus unproductive, motion. They were designed to bring regularity and uniformity out of complexity and reflected a practice that saw in complex and variable human motion a pathology in need of therapy.

The apparatus on display in the exhibit illustrates a managerial attempt to bring people's bodies, intricate and complex, into an industrial order demanding regularity and uniformity. The efficiency apparatus and the efficiency tools of training, practice, and rule making created places where the distinguishing characteristics and features that marked people as autonomous individuals mattered less and less. Workers had instead become elements to be stabilized in the managerial pursuit of economic transformation.

CHAPTER SIX

Pride in Efficiency

The Dispute over *Time on the Cross*

American slavery had been efficient, more so than agriculture in the northern states. So announced Robert William Fogel, future Nobel laureate in economics and professor of economics and history at the universities of Chicago and Rochester, and his coauthor Stanley Engerman, professor of economics and history at Rochester, in their 1974 study *Time on the Cross: The Economics of American Negro Slavery*.[1] They concluded that slaves were good workers and more efficient than laborers on the farms of the North and that their descendants should take pride in this record of "achievement under adversity." What followed was one of the most high-profile academic disputes of the decade, acrimonious and covered by the national press and national television. The book's argument that slaves had been competent workers was passionately embraced by many who were looking for a way to describe the legacy of slavery as something other than mere exploitation; the book was devastatingly criticized by scholars for its assumptions, methodology, and use of evidence.

In the 1989 volume *Without Consent or Contract: The Rise and Fall of American Slavery* Fogel reversed himself, arguing that the efficiency of slavery had arisen not from slaves' good work ethic but from the oppression of gang labor. "Efficiency is not a synonym for good," he wrote, describing a position it had taken him years to develop. His calculation that plantation slavery was more efficient than free farming had undermined his belief that property rights and economic progress were tied to freedom and justice. Fogel had believed that the evils of slavery would manifest themselves in the economics, but the numbers proved him wrong and left him stranded, unable to justify his deep belief in the evils of slavery.[2]

Time on the Cross and *Without Consent or Contract* were attempts by Fogel to resolve the contradiction he perceived between the efficiency of slavery and its

basis in oppression. The dilemma arose because Fogel could not distinguish between a dynamic, transformative efficiency and a static efficiency of balance. He
saw no difference between efficiencies achieved by managers, or plantation owners, who employed rational choice in strategies of growth, and the efficiencies
demanded of those they managed, slaves, who were most valued when they did
not upset the balance and could be harshly punished if they did. He could not
conceive of efficiency through force. Efficiency, in *Time on the Cross,* had been the
by-product of a system in which slaves were diligent and cooperative workers and,
as valuable property, were relatively well treated and judiciously managed. In
Without Consent or Contract the efficiency of slavery came from a different source,
arising not from cooperation and diligence but "directly out of the oppression of
its laborers." The efficiency of slavery came down to gang labor.[3]

Fogel's Dilemma
The Centrality of Efficiency

The original motivation for *Time on the Cross* came from questions and assumptions about efficiency, rather than an interest in slavery. Fogel had not been
particularly interested in slavery at the outset of his career; he, like many cliometricians, saw in it a laboratory for economic technique. "An accident of scholarship," as Fogel put it, "made the economics of slavery a major testing ground for
the application of quantitative methods to the study of history."[4] That accident
was the 1958 publication of Alfred Conrad and John Meyer's economic analysis
of slavery. Conrad and Meyer had argued that an investment in slaves could be
analyzed as though it were an investment in "any other kind of capital"; such an
assessment demonstrated that slavery had been highly profitable.[5]

This was not news to historians of the American South; many had long considered slavery to have been a profitable and flourishing institution, although
some questioned the quantitative methods used in the study.[6] To a new scholar
the situation looked different. Fogel, then a graduate student in economics at
Johns Hopkins University, found Conrad and Meyer's analysis both stimulating
and disturbing, and it raised for the first time, for him, the apparent paradox of a
system simultaneously reprehensible and yet economically sound. The paper em-
broiled most of the faculty and graduate students of his department in a weeks-
long debate and touched off a crusade among younger economists to extend
quantification into the humanities by attempting a self-consciously scientific
interpretation of the slave economy.[7]

This crusade to extend quantification suggested a norm of its own: history practiced scientifically, rigorously mathematical and statistical rather than impressionistic and fragmentary, and dedicated to uncovering general historical laws.[8] This norm underlay all of Fogel's contributions to the slavery debates and was at the root of the indictment of slavery he later offered in *Without Consent or Contract*: there he insisted that even a moral condemnation must account for the economic facts of slavery as practiced. As Fogel later described it in the "Afterword" to *Without Consent or Contract*, he had long viewed slavery as an evil of unconstrained power exercised by slaveholders over their slaves and as cruelly exploitative: rational, yes, but "coldly" so.[9] The rationality was important: it was the basis of an economic view that saw rational choice as the agent of economic transformation. As Fogel portrayed it, this construction replaced a more youthful view, influenced by a boyhood in New York in the late 1930s and 1940s, riding subways decorated with antiracist slogans taken from Ruth Benedict and Gene Weltfish's *The Races of Mankind* (1943), and by his undergraduate years at Columbia, reading Gunnar Myrdal's *An American Dilemma: The Negro Problem and Modern Democracy* (1944) and joining popular student protests against capitalism, fascism, and racism. In his youthful and unclear vision slavery was cruel, the South was backward, and slavery and later segregation were reinforced by bad racist science and businessmen's theories, propounded to help keep labor cheap and disorganized.

The Conrad and Meyer paper of 1958 changed Fogel's views. Conrad and Meyer had made two arguments: that slavery had been quite profitable and had not been dying away and that the slave market had allocated slave labor efficiently, by which they meant that the Old South had served as the breeding ground for slave labor for the cotton plantations of the West.[10] Fogel found it difficult to believe that such a reprehensible system could be so profitable, and he found it especially horrifying to think that slaves had been bred for profit, like animals. Conrad and Meyer's work suggested that the antebellum South had not been economically backward; it also, to Fogel, made the slave system appear even more inhumane than he had thought. Profit and inhumanity coexisted.

But profitability and efficiency were not quite the same thing. One could create an arbitrary measure of the efficiency of investment as a function of profitability, comparing the profits to be had from a capital investment in slaves with other types of investment, for example, as Conrad and Meyer had done. But profitability said little about the processes of production and how they were controlled and managed. That it was profitable said little about how slavery worked. This is where Fogel entered the field.

In seeking to extend Conrad and Meyer's work, Fogel turned to productivity and the related concept of efficiency, rather than profitability. He hoped to reveal something about how slavery had worked, and measurements of productivity and efficiency offered him a way to do that. The two concepts were quite similar. Productivity started with a baseline, often of time—so many widgets produced per hour, and so on. Productivity established a baseline and an output measurement, whereas efficiency analyzed what happened to the factors in the baseline, what happened with the time, with the money, with the labor. Productivity spoke to the outcomes achieved with various inputs; efficiency spoke to how those inputs were used. Behind efficiency lay a legacy of balance and a worry about waste, expressed in its assumptions that one ought to get as much as possible out of what one had put in, not only enough to be productive or to show a profit but enough to show that the system was under control.

Fogel described a great change in his views of slavery between 1968 and 1972, the years in which he and Engerman began to study the relative efficiency of slave and free agriculture. He was disturbed by two findings: that slave plantations had been more efficient than free farms and that slave breeding had been practiced neither systematically nor deliberately for profit. The first of these findings undermined his assumption that slavery had made slaves apathetic and had worn them out before their time, and the second undermined what he had thought of as the worst indictment of slavery, the breeding of people for profit. Together they presented him with a dilemma: how morally to characterize slavery. He had not, at this point, jettisoned the idea that efficiency had a positive moral value. The findings had not yet undermined his belief that efficiency came from free and willing workers. He was not yet to the point of emphasizing slavery's evil and cruelty.

How to characterize oppression was one part of the problem. Productivity in the face of exploitation seemed to defy economic interpretation. Fogel could not see how people "so deeply oppressed," demoralized, and denied incentives and opportunities could outproduce people who were free. His efficiency findings did not match his view of half-starved and listless workers fumbling through the day. On the contrary, they implied, as he put it, "a far greater level of labor discipline and a far greater degree of, if not cooperation, at least acquiescence to the objectives of the planters than I was prepared to entertain." In short he could not see how oppressed and exploited people could produce things of value.[11]

This was not an uncommon dilemma. Stanley Elkins described it in an article for *Commentary:* the problem, he wrote, "is how you can emphasize vitality and achievement while you are also emphasizing decline and degradation." Historians of the Industrial Revolution struggled with it, trying to account for two simultaneous aspects of working-class experience: exploitation that could be brutal and demoralizing, alongside resistance that was energetic and creative and helped to build a distinct working-class culture.[12] Robert Coles noted this dilemma among northern civil rights workers who went South and found it difficult to reconcile the vibrant black culture they encountered with the oppression under which many blacks continued to live.[13]

For Fogel the dilemma had a particularly economic cast. Oppression was not only antithetical to liberty; it was antithetical to economics as he understood it. Fogel's commitment to an economic explanation of the world also included a commitment to one of its principal neoclassical liberal assumptions: that the economy and the market functioned because people made free and informed choices. Oppression, by definition, interfered with choice. The efficiency findings thus required Fogel not only to rethink the economics of slavery but to rethink the oppression as well. Fogel's search for an adequate indictment of slavery illustrates, in his own earlier thinking, an intimate link between economic success and virtue, or between efficiency and morality. He repeatedly described his early efficiency findings as disturbing. Not only were they counterintuitive to him and contrary to his expectations, but they also left him with inadequate grounds to oppose slavery. His was a crisis of faith, of an economic faith, in an economic view of the world.

The Appeal of *Time on the Cross*

Little, Brown published *Time on the Cross* in the first week of May 1974. Fogel and Engerman had both rethought the issue of the economics of slavery, and nearly two years before publishing their preliminary results, they rechecked their data, looking for computational mistakes and searching for conceptual errors in the way they had formulated their questions. They sought new data that might reverse their findings. In the end they went forward with publication, having convinced themselves that their results did indeed have merit and despite their deep discomfort over the finding that slavery had been efficient.

Number crunching was the method behind *Time on the Cross,* and the key comparison Fogel and Engerman wished to make, between overall productivity in the North and South, required number crunching. They were among a group of scholars championing the relatively new and controversial use of economic models and measurements as tools of historical interpretation, feeding increasingly large sets of data into high-speed computers that performed complex calculations that were previously impossible to perform. At least thirty-five scholars had collaborated in the project, which had been funded by $362,300 in grants from the National Science Foundation. The authors and their graduate students collected data, and they also used data generated by other scholars. They chose 1860 as the year for their comparison and relied on manuscript schedules of the U.S. censuses for agriculture, population, and, for the South, slaves. Data for the North came from 21,118 farms in 102 townships, selected randomly. Data for the South came from 5,229 farms in 11 southern states. Data on slave efficiency were extracted from slave values given in probate records, for approximately 80,000 slaves across 8 southern states. Invoices for the selling of 5,000 slaves in New Orleans and U.S. Customs Bureau shipping manifests yielded data on the slave trade. Fogel, Engerman, and Richard Steckel used the records of 30 plantations—from Alabama, Georgia, Louisiana, Mississippi, North Carolina, South Carolina, and Texas—to create data sets on slave demography and plantation life.[14] Fogel and Engerman also drew from well-known sources such as the travel accounts of Frederick Law Olmsted and the diaries of slave owner Bennett Barrow. Details of evidence used and calculations made appeared not in the main volume of *Time on the Cross* but in a separate and more expensive supplement, subtitled *Evidence and Methods*.

Fogel and Engerman argued that historians, especially since World War II, had described slave workers as inferior, even while rejecting overt biological racism. Ulrich B. Phillips's belief in black racial inferiority had influenced his *American Negro Slavery* (1918), which had been for some decades the standard historical work on the topic. Phillips depicted slavery as a paternalistic arrangement to the benefit of both master and slave, offering slaveholders a distinctive way of life and slaves the training to become civilized contributors to society. Phillips's successors, Fogel and Engerman charged, had substituted a circumstantial argument for a racial one, offering "a respectable and acceptable sociological justification" of black inferiority, as occasioned by a system that so repressed slaves as to have undermined their family feeling, their ability to work productively, and their capacity for self-direction and independent judgment.[15]

Fogel and Engerman cited two particular historians as perpetuators of the myth of black incompetence: Kenneth Stampp and Stanley Elkins. Stampp, whose *The Peculiar Institution* (1956) replaced Phillips's *American Negro Slavery* as the standard study, had argued that slaves engaged in "day-to-day resistance" as a way of exercising influence over their circumstances. Elkins, in a study of the psychological effects of bondage, had made a highly controversial comparison of slavery and Nazi concentration camps. Fogel and Engerman charged that both Stampp and Elkins had described slave workers as incompetent—in Stampp's case because they had taken measures of resistance such as shirking their work, mishandling and damaging tools and livestock, running away, and feigning illness; and in Elkins's study by succumbing to the cruelties of a system based on coercion, which had forced them to adopt a distinctive slave personality: the Sambo, submissive but irresponsible, lazy, and childlike.[16] As Fogel and Engerman saw it, such descriptions illustrated slave inefficiency and, in turn, a racist view of black incompetence. In another criticism of Stampp, buried in the companion volume, *Evidence and Methods,* the authors took Stampp to task for emphasizing slavery's deleterious effects on diet, clothing, health, shelter, and family life, rather than economic issues such as barriers to black education and restrictions on slaves' possibilities for economic and professional achievement. By emphasizing such basic issues, Fogel and Engerman charged, Stampp had misdirected the debate.[17] The economic arguments, in their view, were much more compelling.

Time on the Cross was appealing, partly because it was informal, polemical, and clearly written and also because it was unmarked by the sort of scholarly apparatus that many readers find intimidating. No footnotes appeared in the first volume, the one publicized and reviewed in the popular press and which Fogel and Engerman had been at pains to make accessible and relevant. In a review of the book the economist Peter Passell had in fact commented that it read like a case developed by attorneys unwilling to "cloud the jury's mind with ambiguity."[18] Fogel and Engerman's theses were clear, and they offered ten principal corrections to what they called the prevailing "traditional characterization" of slavery's economics. Slavery was not an irrational system, nor was it doomed to die out on its own; it was not incompatible with southern industrialization, nor did the material conditions of life for slaves compare unfavorably with those experienced by free industrial workers. The southern economy, on the eve of the Civil War, was not stagnating, nor did slave owners feel pessimistic about their economic future. Slavery had not destroyed slave families. Slavery had instead been an efficient form of business, indeed 35 percent more efficient than the northern system of free family farms. Slaves were not inept and incompetent but, on the contrary, were "harder-working and more efficient" than their white counterparts. Nor had slaves been as fully exploited as many believed; in fact, the rate of expropriation had been quite low, and Fogel and Engerman argued that "over the course of his lifetime, the typical slave field hand received about 90 percent of the income he produced."[19]

Efficiency and Morality: The Early Positive Reception

Early reactions to the book were overwhelmingly positive, even glowing. Passell, writing for the *New York Times Book Review,* called *Time on the Cross* the most

important book of the decade, and Walter Clemons in *Newsweek* called it "dynamite." Even ten months later, after it became clear that the book contained fundamental methodological errors and misinterpretations, it was still popularly
described as revolutionary and insightful, as a record of "the triumph of the
human spirit over adversity that [blacks] should be proud to have recorded."[20] A
conflation of efficiency and morality was evident in many of the early reviews.
Time on the Cross itself suggested a positive moral value in efficiency, first by equat-
ing it with competence and then by describing its finding of slave competence as
a corrective to racism. Participants in the dispute conflated efficiency and morality both by describing efficiency as a norm, a standard to which people should
aspire, and by seeing in slave efficiency a source of pride for later generations of
black Americans.

What *Time on the Cross* offered, and what excited a passionate defense, was a
view of slaves as efficient and competent workers. Fogel and Engerman claimed
that, in reassessing slaves' historic and continued role in building American society, their work thus offered a point of pride for black workers. The context was the
search for a "usable past." The term was David Brion Davis's, who used it to
describe a desire to find something positive and worthy of honor in African-
American history, something worth building upon, something other than victim-
ization, brutalization, and infantilization.[21] It was a worthwhile project. The
causes of black poverty and welfare dependency had been fiercely debated, especially since the 1965 Moynihan report. The Moynihan report argued that the
problems of the United States's blacks were largely due to the deterioration of the
black family, which contributed to an ongoing cycle of social pathology. It further
portrayed the problem as a legacy of slavery, as the most important illustration of
the silent and hidden influence of three centuries of exploitation. Many saw in
the report the suggestion that blacks were poor and lazy workers or incompetent
fathers. The report appeared to many to have resurrected old and negative
stereotypes.

The norm of efficiency was created through descriptions of inefficiency as
pathological and by characterizing arguments that slaves were inefficient as
racist. One of the earliest popular reviews made this clear, arguing that "the inefficiency of plantations and black labor came as a natural corollary."[22] Walter
Clemons's May 1974 review in *Newsweek*, widely quoted for the phrase "*Time on
the Cross* is dynamite," elided inefficiency with racism by reiterating Fogel and
Engerman's claim that critics of the immorality of slavery had helped perpetuate
the image of the Negro as inefficient, childlike, and demoralized and by immediately afterward quoting the inflammatory statement of abolitionist Cassius Mar-

cellus Clay that Negroes lacked self-reliance and that they could not be helped: "God has made them for the sun and the banana."[23] Critics as well as supporters of *Time on the Cross* invoked efficiency as a norm. In a critical review published that fall, Winthrop Jordan described the norm of efficiency as a white one. "It is downright pathetic," he argued, "to see the authors so blinded by ethnocentrism that they regard finding slaves 'diligent and efficient workers' as pointing to 'the record of black achievement.' As presumed 'friends' of black Americans, the authors invite blacks to a handsome record of emulating the Protestant work ethic of white Americans. With friends like these, who needs enemies?"[24]

Reviews also presented efficiency as a source of pride. Thomas Fleming called Fogel and Engerman's story "a triumph of the human spirit over adversity that [blacks] should be proud to have recorded."[25] Members of the conservative press assessed the book similarly. M. E. Bradford and Jeffrey Hart, both writing in the *National Review,* found slaves' contributions to a "prosperous and stable South" a reason for pride among their descendants and argued that "contemporary blacks therefore have a right to take a good deal of pride" in the new historical description.[26] Reviews for more liberal publications echoed the case for pride. Christopher Dell, writing in the *Nation,* concurred with Fogel and Engerman about the racist implications of much recent scholarship, noting that black nationalists had for some time been irritated by white historians and sociologists portraying blacks as "incompetent workers, businessmen, and heads of families." Dell thought Fogel and Engerman's benign portrayal of slavery was in bad taste, but he welcomed their argument that "slave-bred inefficiency" was not to blame for the economic difficulties of many contemporary blacks. Dell had sympathy for the central questions behind *Time on the Cross:* whether the record of black performance under slavery had indeed been so terrible and whether slaves had really become the worst possible workers, kept in order only by the whip. He saw contemporary significance in the new study, for it demonstrated that the economic problems facing black families were not remnants of habits acquired under slavery but had developed instead under industrial influences from the North in the years after emancipation. Black incompetence was shown to be a myth; to the contrary, southern economic success had been founded on slave efficiency.[27]

The Critics Gather

Although reaction to *Time on the Cross* was initially positive, within months of its publication scholars began to air numerous and serious criticisms in the press, in scholarly journals, and in conference rooms. Conferences were held at the University of Rochester in October 1974 and at the University of South Carolina, Columbia, in November 1974, and a special session was organized at the annual meeting of the American Historical Association in January 1975.[28]

The book had been widely publicized. The week before it was published, Fogel appeared on the *Today Show,* where he admitted that he and Engerman were waving the book "like a red flag" to draw attention to their new approach. It was a contentious appearance in which his fellow guest, the eminent sociologist Kenneth Clark, suggested Fogel's work was really an apologia for slavery. The book was also widely discussed in the popular press. *Newsweek*'s review came out the week the book was published and *Time*'s shortly thereafter. *Time on the Cross* also excited measured condemnation. Critics condemned it for several things: for its tone of easy mastery of southern history and its handy dismissal of thorny historical problems; for the straw figure it erected in the form of the "traditional interpretation" of slavery; and, as scholars reviewed it more carefully, for its many errors of method and interpretation. It also excited passionate attack in response to an attack of its own: the charge that leading historians of slavery were perpetuating a racist view of black incompetence.

The Rochester conference highlighted many of the most important criticisms of *Time on the Cross*. Engerman and Eugene Genovese, chair of the Rochester history department and a leading scholar of slave culture, organized the event, and more than one hundred people attended, including other prominent historians, sociologists, and economists and members of the national media.

Many of the most serious criticisms at Rochester came from other economic historians. Richard Sutch, discussing the book's argument that slaves had been treated relatively well, exposed what fellow historian Thomas Haskell called "blunders so obvious they would make a sophomore blush with shame."[29] The authors had exaggerated the size of slave cabins by 50 percent, had compared the size of those entire cabins to the bedrooms of New York workers' slums, and had concluded that slaves' dwellings compared well with those of "most of New York City's workers" by comparing them with data on the city's most impoverished residents. Herbert Gutman pointed out that Fogel and Engerman's data on the supposed infrequency of whipping was seriously flawed and statistically misrepresented so as to make whipping appear insignificant and that the authors had taken selective and misrepresentative quotations from the diaries of Bennett Barrow, their major source and well known to historians. Paul David and Peter Temin argued that Fogel and Engerman were not even speaking of efficiency in the accepted sense, in which it measured the productive use, exchange, or allocation of resources. Because their index of productivity was sensitive to consumer demand and market value, it would produce an inflated value for goods for which there was greater demand than for goods in lesser demand, even if they had been produced with equal efficiency. Gavin Wright reinforced this argument, illustrating the importance of the market demand for cotton by analyzing how differing ability to specialize in it accounted for the apparently differing productivity of slave and free farms. Many wondered if *Time on the Cross* deserved further serious attention, and Genovese closed the conference by remarking that the book was a failure, albeit a creative one.[30] Peter Kolchin summarized the prevailing view some years later as one of "bemused condescension" for a "bold but now discredited work."[31]

The book was also criticized for failing to provide the scholarly apparatus that would have enabled others to evaluate its claims. The main volume, unmarked by specialist jargon, was also unmarked by references and notes. Some source material and documentation of calculations appeared in its high-priced companion volume, subtitled *Evidence and Methods,* which purported to give the necessary details. *Evidence and Methods* did include a number of technical details, expressed in a specialist's mathematics and jargon, but even fellow adepts found it poorly organized, difficult to decipher, and lacking at crucial points, some of which could only be established by further correspondence with the authors. To some the companion volume illustrated all that was wrong with what Carl Bridenbaugh, during his tenure as president of the American Historical Association, had called the "bitch goddess quantification": it was abstruse, laden with assumptions, and apt to discount anything that could not be counted. It also suggested an elitist disregard for the views of those Fogel and Engerman described as innumerate, traditional narrative historians, whose methods they claimed to have shown to be inadequate, unscientific, and entirely superseded.[32]

Mechanics of the Measurement of Efficiency

Particularly problematic was the efficiency calculation. It was the foundational idea behind the entire project, as Fogel explained on the final day of the Rochester conference, and it had been singled out as the book's most flawed argument. This was the conclusion of Lance Davis, a prominent economic historian from the California Institute of Technology, and it was echoed in the criticisms of Wright, David, Temin, and others. Evsey Domar of MIT, who developed the productivity index used by Fogel and Engerman, had earlier warned that it ought not be used to measure efficiency or even to approximate it.[33]

Time on the Cross did not report measures of individual slave efficiency, nor did the supporting volumes. None of the calculations measured slave efficiency directly. Analysts inferred the superiority of slave workers by extrapolating from a general productivity index comparing slave agriculture and free farming. Fogel and Engerman's conclusions about the superiority of slave workers came, in other words, not from the numbers but from their interpretation. That interpretation was heavily influenced by Fogel's early assumption that efficiency could not coexist with oppression, an assumption borne out by the benign character of slavery as portrayed in *Time on the Cross* and by Fogel's statement that the early efficiency findings had prompted an outpouring of supporting work on slave culture as well as slave labor and family life. It is the interpretation of slave efficiency that concerns us here and how the vast body of research undertaken to support the efficiency findings—research on the annual slave labor cycle; on diet, nutrition, and infant mortality; on slave culture and family life—ended up leaving the efficiency finding behind. How had Fogel, while never altering the figure of 35 percent, nonetheless greatly reduced his efficiency claim? Efficiency went from representing slave achievement to representing slave oppression.

Fogel and Engerman had used a geometric index of total factor productivity in determining the relative efficiency of southern plantation agriculture. The index contained four terms: output, measured in dollars; and three inputs, capital, land, and labor. Our purpose here is to examine the labor input, its connection to Fogel and Engerman's claims for the superiority of slave labor, and Fogel's later claim for the efficiency of gang labor. Scholars expert in history by the numbers have critiqued the efficiency calculations; our point is instead to consider how the discussion of labor efficiency, and especially gang labor efficiency, resonated with the themes developed here: a split between efficiency for the manager and efficiency for the managed; an important role for surveillance and active managerial intervention; and the necessity of discipline, mechanical or physical and internalized.

Eugene Genovese had taken up Conrad and Meyer's study of slavery's profitability before Fogel and had laid out a program of research into the relative efficiency of southern slave agriculture in the second chapter of his 1965 book, *Political Economy of Slavery*. Genovese argued that measurements of efficiency and productivity could not be done. One might measure profitability, but measuring efficiency and productivity was something else. At its best such a measurement would be cumbersome and unwieldy, filled with a succession of terms relative to other terms and to each other; his example was Conrad and Meyer's measurement of "crop value per hand per dollar of slave price." Assessing productivity would require accounting for a number of things that Genovese thought would be difficult to measure or even to estimate: how much white farmers without slaves, or working alongside slaves, contributed to the southern agricultural product; whether slaves raised the same proportions of the various crops in different years; and how to account for food grown for use on farms or plantations. He was confident that "the plantation was by no means more efficient than the family farm operating in the capitalist economy of the free states" because it did not provide the incentives family farming did and because it had retarded the technological development that made it possible to produce commodities on a large scale. Genovese further argued that southern plantations had not taken advantage of the possibilities of division of labor open to larger enterprises. Planters had few skilled workmen among their slaves and spent significant amounts to purchase northern clothing and shoes as well as to hire blacksmiths and to have doors framed, carriages repaired, and roads built. Plantation slaves often worked with antiquated tools and technologies, perhaps heavy enough to prevent breakage but too heavy for efficient use, or eschewed advances such as the cultivator, which many slave owners decided was too delicate to entrust to slaves, even though it had become standard equipment on northern farms.[34] A measure of efficiency would be extraordinarily difficult to achieve, Genovese believed.

Fogel took it on nonetheless. The index he and Engerman used in *Time on the Cross* took the form of a classic efficiency measurement: a ratio of output to input, with the North's efficiency arbitrarily set at 100 percent. Measures of things in the North were in the denominator, on the bottom, in each of the four ratios used, one of output and three of input: land, labor, and capital. Each term was figured on its own, in equivalent units that canceled each other out, leaving as the result a dimensionless number. The need to express a wide variety of inputs—and of products, or outputs—in equivalent units required Fogel and Engerman to make a number of assumptions and calculations that were later criticized; Thomas Haskell called the whole thing a pyramid of assumptions. For example, Fogel and Engerman gave the same weights to each of the three inputs, of land, labor, and capital, for both the North and South, even though they acknowledged that the structures of the two economies probably differed. They derived those weights from their own estimated production functions for southern agriculture, which indicated that owners of labor received 58 percent of the total product, owners of capital 17 percent, and owners of land 25 percent.[35] Critics David and Temin later concluded that the failure to account for regional differences in prices biased the calculation in favor of the South.[36]

Labor was measured in man-years, or the number of hours a man in his prime would have worked in one year. This required a prior series of calculations in order to set the man-year standard and to convert the labor of women, children and adolescents, and the elderly into equivalent man-years. The value of capital and land was measured in dollars. These, then, were the basic units of the measurement: dollars and man-years.

Critics of the Measurement

Critics analyzed all aspects of the efficiency calculation, from its estimates of the value of land and capital to the role of inflated cotton prices in the measure of output. Fogel and Engerman replied. They did not substantially revise their efficiency measure in subsequent publications, but they, and especially Fogel, did change their interpretations. Fogel did not, however, come to grips with one key criticism: that the very basis of his economic analysis had been misapplied. He did not answer charges that, at its foundations, the economics he had used privileged masters over slaves.

Paul David and Peter Temin made the most sustained critique of the moral implications of available economic methods, beginning with a contribution to the Rochester conference. They argued that neoclassical economics did not offer adequate grounds for the analysis of repressive societies, for it was founded on assumptions that did not pertain. At the heart of the neoclassical welfare economics within which Fogel and Engerman worked was the belief that people maximized their utility by making informed choices in the marketplace. Slaves were only rarely in a position to do this. And although economies or societies could be compared on the basis of how closely they approached the optimal allocation of their resources—on the basis of what economists termed their "welfare efficiency"—this measure was itself only apparently neutral. David and Temin pointed out that it privileged the owner's point of view because the owner was the one in a position to express legitimate choices through market behavior. Slaves were not. "The natural predisposition of economists is to consider how affairs appear to the human agents who are capable of acting so as to better satisfy their wants," they argued. "Modern economic welfare theory simply does not encourage one to view life also from the standpoint of the lathes and the livestock."[37]

This point became especially clear when considering Pareto efficiency, a test of optimality developed by the Italian engineer-turned-economist Vilfredo Pareto. Allocation of resources is Pareto efficient when it is impossible to make a change for the benefit of one person without harming another, or, in other words, when no one can be made better off without simultaneously making someone else worse off. Pareto analysis sets up a status quo and measures changes against it, allowing changes only by unanimous consent. Without such consent it presumes that the system is already Pareto-efficient. The burden is on the change: it must be demonstrated to do no harm and not merely to balance new harms by newly gained benefits. It is difficult to imagine a scenario in which slaves would be able to compensate their masters sufficiently to convince those masters to agree to emancipation. Slaves could be analyzed like livestock or other such investments, David and Temin charged, but no economic model existed to account for a legitimate desire for freedom and to treat it alongside the desire at the core of economic theory: the desire to maximize profits. David and Temin's criticism suggests that *Time on the Cross* measured not slavery's efficiency but that of the slave masters. The problem is that masters' efficiency was not a question of a work ethic but of effective enforcement of a master's will.

Criticism stimulated further research. Cliometric studies of slavery continued, including work on slave nutrition, health, and height; on agricultural work schedules in the North and South; on slave fertility and mortality; and on slave quality of life. Three principal volumes of criticism of *Time on the Cross* appeared, followed by an exchange in the *American Economic Review* in the late 1970s and early 1980s. A number of scholars reconsidered particular aspects of *Time on the Cross* as they bore on their own research projects.[38] Between 1989 and 1992 Fogel published *Without Consent or Contract* and three volumes of supporting materials; he then declared an end to the slavery phase of his career and consigned his slavery papers to the Special Collections at the Libraries of the University of Chicago. Fogel continues to characterize the dispute as "a debate in which there were no losers."[39]

For fifteen years Fogel continued to argue that southern plantation agriculture had been more efficient than northern free farming, and he continued to cite the same figure: by 35 percent. This measure figured prominently not only in *Time on the Cross* but also in *Without Consent or Contract*. He again characterized the efficiency measure as a key finding when he gave the prestigious Walter Lynwood Fleming Lectures in Southern History at Louisiana State University in 2001. His later argument differed significantly but subtly from the one he and Engerman had offered in *Time on the Cross:* the relative efficiency of slave agriculture no longer came down to the superiority of slave laborers but to the gang system and to the more intense exploitation of labor it made possible.[40]

Without Consent or Contract: Fogel Revises His Views

Although Fogel characterized the slavery debates as having no losers, the efficiency rallying cry was one of them. The cheer died away as it became clear that in terms of slavery the roots of efficiency lay in oppression and force. Fogel stuck by the figure of 35 percent, and in fact he later suggested that the figure was probably higher, but by 1989, when *Without Consent or Contract* appeared, he had ceased to offer it as a point of black pride. This stance differed markedly from his position in the early- to mid-1970s, when he was writing and then debating *Time on the Cross*. The difference is revealing and says much about the economic commitments behind efficiency that Fogel found himself forced to reconsider. Fogel's final word on slavery speaks of a confrontation between the assumptions of liberal economics and the human experience of oppression, and of a confrontation between the presumed value of efficiency as a measurement and the types of control that made efficiency possible in practice. Fogel was not the first to see such contradictions, but his struggle shows how deeply rooted the economic commitment to efficiency was. In jettisoning a belief in the moral value of efficiency, Fogel found he had also jettisoned his indictment of oppression and his way of justifying the Civil War.

The Efficiency of Force and Gang Labor

What was gang labor? In *Time on the Cross* it was one of two ways slaves might have been organized for efficient labor. Gang labor was work performed in coordinated gangs, in which tasks were subdivided and performance closely observed. It was characteristic of larger plantations, with fifteen slaves or more, and was particularly suitable for cultivation of tobacco, rice, hemp, and cotton. Members of a gang performed different tasks but in sequence and following a common rhythm; a planting gang might consist of five slaves working in a fixed procession, led by a plowman, then a harrower, followed by a hole driller, a seed dropper, and a raker, who filled and covered the hole. Fogel and Engerman described three ways of maintaining the intensity of the work: by putting fast and strong workers at the front of the gang, to set a good pace; by placing least able workers in the middle; and by "assigning drivers or foremen who exhorted the leaders, threatened the laggards, and did whatever was necessary" to keep up the intensity.[41]

Gang labor, in *Time on the Cross*, supported Fogel and Engerman's argument that southern planters had been effective and rational managers, seeking to maximize their profits, especially in their dealings with the labor force. As they put it, "No question was treated with more gravity than that of labor management."[42] Although individual planters differed in their opinions on how slaves should be housed, fed, and punished, they agreed on the purpose of slave management: to create "a highly disciplined, highly specialized, and well-coordinated labor force," one that Fogel and Engerman described as "rigidly organized as in a factory."[43] Frederick Law Olmsted's observations of the mechanical quality of slave laborers reinforced this conclusion; he described observing a hoe gang of about two hundred workers moving across a field with precision and in parallel lines, too attentive to their work to notice as he loped up on his horse. He found it painful to watch the "stupid, plodding, machine-like manner" in which they worked.[44]

Slave labor may have been the key to the efficiency of large-scale plantations, Fogel and Engerman noted. At least slave owners thought it was. Fogel and Engerman presented gang labor as one type in a mix of positive and negative incentives that planters used to manage their slaves. Labor might also be organized according to a task system, in which a slave was motivated to complete an assigned task by being given leisure time once it was done, time to work his or her own vegetable plot or to accomplish other chosen tasks. Left open was what the authors called the "imperative question" of whether the efficiency of southern plantations came from superior management or from the responsiveness of the slaves. They discussed a variety of motivating tools, including gifts and cash bonuses and the right to use plots of land or to own or sell small livestock. Gang labor was described as a form of scientific personnel management, training black slaves in industrial rhythms of work. It described the regimen as "more like a modern assembly line" than were many antebellum manufactories.[45]

The gang system shared with scientific management an emphasis on continuing and active surveillance. Fogel and Engerman cited Olmsted on the management of gangs: Olmsted had observed drivers following the most minute actions of their gangs, delimiting their specific tasks, and exerting authority over them even in off-hours. A driver "is made to do the duties of policeman, and even of police magistrate," Olmsted noted, and was allowed to whip the members of the gang to keep them in order.[46]

Participants in the slavery debates continued to emphasize surveillance as a feature of gang labor. Ira Berlin described the minute inspection of labor that accompanied early tobacco farming,[47] and Sidney Mintz, discussing variants on Jamaican plantations, noted that strict and continual supervision was central to gang labor.[48] Gutman and Sutch, in their critique of *Time on the Cross*, quoted Bennett Barrow's diary, which describes such surveillance in detail: "am directing them to make a slow and sure lick in one place and to cut the full width of the hoe every time—unless reminded of it they would stand & make 4 or 5 licks in one place, tire themselves & do no work, have several grown ones that work harder & do less than any in the field."[49] In the later book, *Without Consent or Contract*,

Fogel cited observer Robert Russell, who described gang labor discipline as militaristic. This quality made possible a uniform regularity, which Barrow saw as necessary to successful plantation management. "A plantation might be considered as a piece of machinery," Barrow wrote. "To operate successfully, all its parts should be uniform and exact, and its impelling force regular and steady."[50] Fogel described these efforts as time and motion study, yielding a system with "an assembly-line type of pressure."[51]

Force underlay gang labor. Whatever efficiencies it made possible, they came from a physical form of discipline that was analogous to the mechanical form that helped extend machine efficiencies. The question of force gave Fogel and Engerman a great deal of trouble and accounted for some of the most embarrassing criticisms. They had acknowledged that force was the only way to get blacks to accept gang labor but argued that owners and overseers had used it judiciously, "not cruelly, but optimally."[52] Their view of optimizing force had been based largely on Barrow's diaries, well known to historians for the record of whippings they contained. Fogel and Engerman had calculated an average of 0.7 whippings per slave during the twenty-three-month period covered by the diaries. Gutman and Sutch argued that this translated into 70 percent of the women and 80 percent of the men having been whipped at least once during that time and that 73 percent of the whippings were given for poor labor: the more productive pickers had in fact been punished the most. Kenneth Stampp had cited Barrow's report of indulging in a "general whipping frollick," of whipping every hand in the field, including the driver, and of breaking his cane over the head of one slave and cutting another badly with a club.[53] Drivers also wielded force, as Gutman and Sutch illustrated with Olmsted's description of a slave driver, "a very tall and powerful negro," walking alongside the gang, cracking his whip, and shouting surly reprimands.[54]

By 1989 Fogel had concluded that the measure of slave efficiency came down to exploitation and oppression. Slavery, he argued, was evil because "its productive efficiency arose directly out of the oppression of its laborers." He quoted David Brion Davis, a critic of his earlier work, to substantiate the point. The slave system "transformed each slave into 'the mere extension of his master's physical nature' and will" and required an "extreme degree of domination."[55]

That the measure of slave efficiency came down to gang labor demonstrates how important force had become among the developing features of industrial efficiency. The point here is not that force made slaves efficient but that the search for efficiency turned up a system of force. If efficiency requires mechanical or physical discipline, practiced by an active and intervening agent in pursuit of a goal, and if it requires systematic management and surveillance, then the place to find it would be in a system of domination and exploitation. The more perfect the domination, the more perfect the exploitation, the more efficient the system would be.

It would be so because efficiency was designed to minimize resistance. It was created to keep things from getting in its way. Resistance is antithetical to efficiency, and one of efficiency's historical features has been its negative function in its elimination. The extent to which slaves resisted the system that held them in bondage was hotly disputed in the slavery debates. Fogel and Engerman discounted resistance, saying that there was little of it and that it was ineffective. They offered this argument as a corrective to Stampp's argument that slaves had engaged in day-to-day resistance, thus creating a severe discipline problem for their masters. The debate over resistance underscores the importance of control: the less that slaves resisted and the lower their chances of influencing the system through such resistance, the greater the control to which they were subject. Michel Foucault, foremost theorist of relationships of power, domination, and control, made a subtle distinction between localized power relationships and systems of domination. Power relationships were characterized by negotiations over authority and by mutual accommodations between those subject to power and those who wielded it. Domination was characterized by structures of power that had congealed and become largely impervious to resistance.[56] Fogel and Engerman's argument that slave resistance was scattered and ineffective demonstrates not that slaves willingly submitted but that the domination under which they lived had so congealed that their efforts could do little to dislodge it or to open it up to reconsideration.

The "Afterword": A Statement of the Neutrality of Efficiency

Fogel's concern about efficiency underlay the key issues in the "Afterword" to *Without Consent or Contract:* how to characterize oppression, when economics was not enough; how to characterize efficiency gained by masters at the expense of their slaves; and how to relate historical facts to contemporary moral positions. Efficiency, as the "Afterword" illustrates, was part of an economic commitment deeper than Fogel's worries about slavery and oppression. Fogel struggled with the problem of efficiency and oppression, but he did not address the most foundational criticisms of the measurements he had taken up. In particular, he failed to consider the implications of an efficiency gained not in spite of oppression but through it. He had measured efficiency that accrued to masters at the expense of their slaves, and he remained committed to an economic analysis.

Fogel's dilemma remained unsolved. Arguing that slavery was not only profitable but efficient appears to have been painful to him, although he did not say so; the problem occupied him for two decades, from the late 1960s to the late 1980s, as he found himself bitterly attacked as a defender of slavery and went to great lengths, especially in the "Afterword," to establish his antislavery credentials. The apparent contradiction between efficiency and oppression that Fogel described in the "Afterword" was for him a real and festering sore. Early in the "Afterword" he cited a short piece in the accompanying volume, *Evidence and Methods,* on how moral issues affected those involved in the slavery debates. He described the discomfort and embarrassment felt by many trying to quantify the business of slavery, who found themselves measuring rates of return on purchases of children or efficiency indexes of things they thought immoral. As Fogel saw it, few escaped feeling personally sullied by the dirtiness of the system they studied, feeling that something of it rubbed off on themselves, just by scholarly proximity. At the end of the piece he quoted a letter from John Meyer, coauthor of the 1958 paper that had so influenced Fogel and his colleagues. Meyer had written of the personal cost felt by those continually attacked as heartless or insensitive to suffering because of their work with the numbers and how he and Conrad had found it "thoroughly upsetting," not only because the attacks were personal but because they were often fairly substantive. It was the substantive nature of the dilemma that plagued Fogel, not that he was called names.[57]

Fogel's dilemma was substantive because it had not only to do with efficiency and oppression but also with economics and oppression. Efficiency was not the whole problem; it was, for Fogel, the route to a broader consideration of morality and economics. In the "Afterword" discussion of efficiency took Fogel directly to considerations of oppression and the nature of property rights, economics, and government. His consideration of oppression led him in turn to decouple efficiency and virtue.

To critics, however, it was not enough to decouple efficiency and virtue. Doing so left open to question the whole economic apparatus that efficiency had justified. Peter Kolchin, reviewing *Without Consent or Contract,* remarked that it was subject to the same criticisms of approach and methodology as *Time on the Cross.*[58] Harold Woodman shared this judgment; Fogel had still to come to grips with the implications of his economic method, Woodman wrote in 1995, for he continued to treat the absence of free labor as but a minor quirk in a southern form of capitalism, rather than an important and perhaps crucial feature in how the southern economy should be defined.[59]

Ralph Austen put his finger on the inadequacy of Fogel's middle-of-the-road offering in a review of *Without Consent or Contract* for *Business History Review.* Austen saw in Fogel's new position a retreat from provocation and an embrace of "a mainstream liberal position," seeking balance between the benefits of a free market and the need to provide social benefits that the market cannot. Reasonable, yes, but according to Austen inadequate, for it failed to address the ideological contradictions it comfortably covered over. The nature of slavery as an economic institution must be squarely confronted, Austen believed, and not merely balanced by discussion of its "human" side. If slavery was profitable and efficient, and also a form of liberal capitalism rather than an aberration of it, then the categories of liberalism itself, and of profit and efficiency, must also be questioned and not merely balanced.[60]

Some historians saw the interconnection of morality and economics as characteristic of slavery itself and not just of its analysis. Davis writes of the "profound contradiction of a free society . . . made possible by black slave labor" and of the ties between profits and a culture of liberty, especially a modernizing culture promoting individual entrepreneurship and profit while simultaneously creating a moral and ideological system to justify it.[61] That system of justification, especially its enthroning of profits as the measure of success or failure, is what gave the moral tone of virtue to Fogel's discussion of efficiency. Even though by 1989 Fogel could clearly state that he had jettisoned any belief in the inherent virtues of efficiency or economic success, he turned again to economics to buttress his indictment of slavery and to justify the Civil War. He described the northern victory in the war as "econocide."[62]

The assumption that efficiency was achieved only through rational choices made by free and willing agents was at the root of Fogel's dilemma. He conceived only of a dynamic efficiency of transformation, aimed at growth and rooted in calculated management. He could not conceive of efficiency through force, or of an efficiency of stasis or balance, in which what mattered was not dynamic transformation but keeping in line and doing what was expected. Fogel's discovery that gang labor was at the root of slavery's efficiency demolished his assumption and led him to a strenuous revision of his own earlier position.

Fogel's final word on slavery, in the "Afterword" to *Without Consent or Contract,* speaks of a confrontation between the assumptions of liberal economics and the human experience of oppression and between the presumed value of efficiency as a measurement and the types of control that made efficiency possible in practice. He retained a belief in the importance of rationality and management, but he let go of the idea that efficiency rested on willful and free individual initiative. It was not an easy abandonment, and the reaction to his "Afterword" suggests, despite scholars' protests to the contrary, that many people continue to believe efficiency to be good, on its own and in and of itself.[63]

CHAPTER SEVEN

Global Efficiency

An Enduring Industrial Value in a Postindustrial World

The assembly line of the early twentieth century is the most potent symbol of modern industry, associated with the automobile plants of Henry Ford and immortalized by Charlie Chaplin in the film *Modern Times* (1936). The assembly line epitomized how central was the flow of production, the ongoing, uninterrupted chain of planned and choreographed motions. With its conveyor belt and continuous operations, it "eliminated challenges and surprises as the product moved along with every step worked out beforehand."[1] Flow was as crucial to conceptions of efficiency as it was to the assembly line, flow of water, flow of work; efficiency's role was to manage or prevent disruptions to the stream or current.

Flow is also a celebrated feature of globalization, in the instantaneous flow of information between all but the most remote places and the near-instantaneous flow of capital. Descriptions of globalization range from the clear and functional to the complex and experiential, from Joseph Stiglitz's "removal of barriers to free trade and the closer integration of national economies" to Steven Best and Douglas Kellner: "flow of capital, goods, information, culture and entertainment, and people across a new networked economy, society, and culture . . . an overdetermined and complex phenomenon."[2] The phenomenon is marked by ceaseless movement and the crossing and recrossing of borders. Globalization is a "space of flows."[3]

Efficiency is an industrial value, having been developed to assess the performance of the great machines that powered industrialization: waterwheels, steam machines, internal combustion engines. It seems out of place in a postindustrial world, which emphasizes the plurality of values and is girded underneath by the new global economy, built on networks rather than hierarchies and dependent on flexible organization rather than centralized planning. This chapter asks what role efficiency can have in a globalized, postindustrial world. It is not historical but contemporary and examines three areas in which efficiency encounters flow: in American jurisprudence, in the legal recognition of administrative efficiency as an important value; in calls for early childhood education to prepare a workforce that can compete globally, in which efficiency is seen as also promoting social equity; and in the computing industry, particularly in tools used in process industries, in which the image of flow is most explicit. Opposed to flow is disruption; efficiency appears, in these examples, as a way to prevent disruption or interference.

A variety of theories of the postindustrial age have been developed, many between the late 1960s and the mid-1970s, as social observers noted changes in patterns of waged work. The most influential came from the sociologist Daniel Bell and was part of his larger analysis of changes in, and the passing of, modern society. Bell emphasized a shift from manufacturing employment to work in the services, along with the increasing importance of knowledge work and hence formal education instead of training on-the-job. Bell thus predicted the terms in which the economies of the wealthiest regions are now routinely portrayed. Bell's description allows an analysis of efficiency, flow, and disruption in a period of changing, and globalizing, work.[4]

At issue is the interplay of freedom and constraint. The flexibility made possible by telecommunications and information technology is commonly seen as evidence of a newfound freedom, for both business and workers, who no longer need be constrained by geographical locations or standard work hours. The envisioned freedom is from the demands of typical, modern industrial production, with a highly capitalized industrial plant and standardized work and task schedules. Efficiency is strongly associated with the traditional industrial process, and one of the appeals of a postindustrial society is the freedom it promises from efficiency's demands. Manufacturing is increasingly automated, leaving workers free to enter the service sector or the knowledge economy, in which efficiency concerns not so much physical gesture as a more general sense of productivity. But the surface flow that characterizes the globalizing economy, and which gives the illusion of great freedom, is supported by rigid structures. Legal systems undergird markets; financial regulations direct the flow of capital. Even the famously free market of Adam Smith, balanced by an "invisible hand," depended upon a set of restrictive conditions.[5] Telecommunications themselves, including the Internet, rely upon precise protocols and standardized components in order to function consistently; historically, computing relied heavily on formal, mechanical models of closed systems designed to eliminate uncertainty.[6] These structures, and the coordination upon which they depend, remain susceptible to interference. This is where efficiency can continue to play a role.

Efficiency, Disruption, and the Law

The connection between efficiency and disruption within a traditional national context is made explicit in American law, in which efficiency has been an important and contested concept. Legal considerations of efficiency make clear its value as a tool of administration or management. Efficiency describes measures designed to streamline the settling of lawsuits, helping to manage increasing burdens on the legal system. It has more potent and particularly legal meanings in constitutional law, in which it is recognized as an important interest to be balanced against claims such as free speech and among jurists who see efficiency as itself the embodiment of justice. Both efficiency's practical and administrative use and its normative role in determining judicial or legal action are disputed.

"Truth becomes a hostage to the efficiencies gained from negotiated pleas" is Thane Rosenbaum's *cri de coeur* against a heartless legal system in which "bureaucratic efficiencies trample all other values."[7] Rosenbaum, lawyer, novelist, and legal educator, sees efficiency as the "cold bottom-line" to American justice, arguing that the administrative reliance on efficiency makes it impossible to achieve meaningful results within the U.S. legal system.[8] The efficiency of settlements and plea bargains comes at too high a price, in Rosenbaum's view: it denies victims the opportunity to tell their stories, in public and in full. The system spreads discontent instead of settling it and often suppresses facts rather than revealing them. The insistence on order and decorum in the courtroom may serve efficiency, but it is a managerial tool at odds with justice itself, Rosenbaum charges; it helps to preserve a veneer of impartiality and objectivity by stifling expressions of deep and complex human emotion. Law as practiced relies heavily on stipulated common denominators, principally money and jail time, and reduces injuries to questions of financial or bodily harm and redress to monetary damages or incarceration. Such common denominators serve efficiency by making it possible to settle disputes with the least expense of time, money, or effort, and by making immaterial issues that do not lend themselves to ready quantification in these terms. Rosenbaum describes such efficiency as a response to the scarcity of resources and time in the court system and charges that such procedural measures, while expediting the flow of cases through the system, have altered the substance of the law.

U.S. courts have themselves recognized administrative efficiency as an important value, at times important enough to offset other and cherished rights, and opinions from the bench on the balance between efficiency and constitutional freedoms have become part of the common law. An example comes from a series of cases concerning the constitutional freedom of speech for government employees. In general a government employee may be disciplined or dismissed for speech that impairs the efficiency of the agency for which he or she works, if that speech does not also pertain to "any matter of political, social, or other concern to the community."[9] This from the U.S. Supreme Court in *Connick v. Myers* (1983), a case that continues to be cited in decisions in which governmental efficiency conflicts with employees' freedom of speech. The phrasing in *Connick* itself stems from a 1968 case, *Pickering v. Board of Education.*[10] *Connick* and *Pickering* treated efficiency as synonymous with management's ability to achieve its own specified outcomes. In neither case was efficiency measured; rather, it was intuited as a function of the employer's good judgment. The rulings speak of maintaining discipline and avoiding disruptions and have been influential in considerations of academic freedom of speech.[11]

Connick concerned the case of Sheila Myers, an assistant district attorney for the Parish of Orleans, Louisiana, who was fired for circulating among her colleagues a questionnaire about employee morale and personnel policies in the Office of the District Attorney. She had worked there for more than five years, competently prosecuting criminal cases, when she was informed that she was to be transferred to cases in different sections of the criminal court, a transfer that she opposed. Myers was outspoken in letting her employers know of her opposition, and when she learned that she was to be transferred anyway, she compiled a questionnaire asking her fellow employees what they thought of the transfer policy, about the general level of morale in the office, whether they had confidence in their supervisors, and if they had ever felt pressured to work on a political campaign. Harry Connick, the district attorney, fired her for refusing to accept the transfer and called her distribution of the questionnaire an act of insubordination. Myers sued, arguing that her constitutionally protected right to freedom of speech had been violated. The case came to the Supreme Court after a U.S. District Court and the Fifth Circuit Court of Appeals had agreed with Myers, having been persuaded that the real reason for her dismissal was the questionnaire, rather than her objection to the transfer.

The Supreme Court reversed those decisions in a five-to-four vote, the majority agreeing with Justice Byron White's opinion that Myers had been speaking on a matter of personal interest rather than public concern and that her employer's belief that she might disrupt the efficient functioning of the office outweighed Myers's free speech rights in the one item in her questionnaire that might have been of public interest, whether assistant district attorneys had felt pressured to work on political campaigns. The dissenting justices affirmed the finding of the district court from which the case had come. Justice William Brennan, among those dissenting, argued that speech about how government is or should be run was indeed of public interest and that Connick had failed to show any evidence that Myers's actions had diminished the efficient functioning of the office.

The efficiency argument came from *Pickering,* from the wording of a Will County, Illinois, school board report on the effects of a teacher's letter to the editor of the local paper. Marvin Pickering taught in Township High School District 205 of Will County, and in a letter to Graphic Newspapers, published in September 1964, he criticized the recent handling of revenue-raising proposals by the school board and the superintendent of schools. The school board dismissed him, after determining, in its words, that his letter was "detrimental to the efficient operation and administration" of the schools in its jurisdiction. The Circuit Court of Will County upheld his dismissal, as did the Illinois Supreme Court; the U.S. Supreme Court then reversed the lower courts' decision, all nine justices concurring in whole or in part with Justice Thurgood Marshall's opinion. Justice White also dissented in part.[12] At issue in *Pickering* was the relationship between efficiency, discipline, and controversy. Efficiency would be undermined because faculty discipline would be undermined, the board charged, and because the letter might occasion "controversy, conflict, and dissension," not only among teachers but also among board members and citizens. Much of Pickering's letter was false, the board argued, and damaged its credibility and the reputations of its members.

In *Pickering,* as Marshall put it, the Court found itself called to balance "the interests of the teacher, as a citizen," who ought to be free to comment on matters of public concern, and education was certainly one, "and the interest of the State, as an employer, in promoting the efficiency of the public services it performs through its employees."[13] In making that balance, the Court decided that Pickering had indeed spoken to a matter of public concern; that his statements had been largely correct, although he had made honest mistakes on a couple of points; and that the school board had given no evidence that the letter had impaired Pickering's performance as a teacher or the functioning of school administration—that is, it had not impaired efficiency. How the Court might have weighed Pickering's speech rights had there been some impairment of efficiency is not clear, although that is exactly the balance Marshall called for it to make. The call for balance influenced a number of later cases, including *Connick,* in which White wrote of "the *Pickering* balance."[14]

Pickering and *Connick* illustrate how efficiency is legitimated as necessary in the pursuit of organizational goals. They deal specifically with governmental functions but reveal more generally how management for efficiency requires not only positive, productive measures but also the removal or prevention of potential disruptions.

Efficiency, Equity, and Early Childhood Education

In economic policy a long-recognized difficulty has been the perceived trade-off between efficiency and equity, between how resources are distributed by the market (efficiency) and how they might be distributed as a matter of fairness and social justice (equity).[15] It was the central issue for economist and Nobel laureate James J. Heckman when he addressed a forum on early childhood education, convened in New York, in January 2006, by the Committee for Economic Development, the Pew Charitable Trusts, and the PNC Financial Services Group. Heckman argued that "investing in disadvantaged young children is an economically very efficient policy," in addition to being equitable.[16] Heckman thus summarized a growing movement among businesses to support early childhood education measures. Many see investment in programs aimed at the youngest children as an especially productive way to guarantee the availability of a well-trained and stable workforce while simultaneously helping to address the social problems of poverty and crime. The context of their interest is increasing competition within the globalized economy and worries about the "sputtering" of the American workforce, specifically a stalling in the percentage of people attending and graduating from college, a rising rate of high school dropouts, and a deterioration in family environments, all of which perpetuate the cycle of impoverishment by failing to provide children with adequate stimulation, in both cognitive and other forms.[17]

The point of the campaign for early childhood education is to integrate children into an economy that demands ever-increasing levels of skill and education; if they are truly integrated, they cease to cause disruptions. The benefits of early childhood education are discussed in business and social terms, not in terms of individual well-being, although several studies suggest that participants derive important personal advantages from it later in life. The case has been developed in technical position papers and in conferences and forums in Minneapolis, New York, and Washington, D.C. Meetings and research are sponsored by organizations such as the Committee for Economic Development that aim to give business a greater role in setting social policy.[18] Since the 1950s ideologies of business management have increasingly taken society, rather than simply the firm or industry, as their province. Education itself has long been a tool of social reform and, during the Progressive era in the United States, was also evaluated in terms of social efficiency.[19]

A leading proponent of business investment in early childhood development has been the Federal Reserve Bank of Minneapolis. In the fall of 2003 it hosted a meeting of economists and business leaders to propose a joint government–business endowment to provide disadvantaged children with scholarships to preschools; business leaders across the nation have also become interested in the Minnesota proposal. Educating the littlest ones, and their parents, is fundamentally a question of managing human capital, as it was put by those addressing the meeting. The efficiency calculation is this: every dollar spent on early childhood education produces a real return, perhaps as high as 12 percent, by helping more people to get diplomas and making them more employable and thus cutting crime and raising standards of living. In particular, early childhood education increases the efficiency of later education, measured in better academic performance and in higher earnings later in life. It produces more skilled and productive employees.[20] The *Minneapolis–St. Paul StarTribune* editorialized about the conference: "Pre-K education does more than any other public investment to assure Minnesota businesses that their best edge on the competition—a highly productive workforce—will hold into the 21st century."[21]

Putting early childhood education on the business agenda, as the Minneapolis Federal Reserve meeting did, shows the continuing reach of efficiency. Students who go through such programs achieve a modest return themselves, through higher earnings and a better quality of life—this is the equity payoff— but the real efficiency lies in the creation of a better business environment, with a better, more stable workforce, lowered crime rates, and fewer expenses for public welfare and incarceration. As Art Rolnick and Rob Grunewald of the Minneapolis Fed put it, early childhood education offers an "extraordinary" return on monies invested, a return that cannot be matched by public subsidies to private businesses or funding for sports stadiums or incentive packages to bring corporate headquarters to town. The public return is an efficiency of growth and transformation; the underlying efficiency is more prosaic: fewer students disrupting classes and more of them finishing school.

The common denominator is money. Rolnick and Grunewald transposed students' grades, crime and graduation rates, and classroom disruptions into dollar figures by extrapolating from the cost-benefit analysis of the well-known High/Scope study of the Perry Preschool of Ypsilanti, Michigan.[22] The High/Scope study began in the mid-1960s and followed 123 black children from low-income families through an early childhood program at the preschool, where they spent two and a half hours in the classroom a day and received a one-and-a-half-hour home visit, with their mothers, each week. The Perry Preschool children were compared with a control group who were not part of the program, and interviews were conducted with 117 of the original participants when they reached the age of twenty-seven. The program's teachers were certified in elementary education and in early childhood and special needs education, earned salaries 10 percent above the local school district's scale, and had a manageable workload of an average six students each.[23] Extrapolating from Perry Preschool data, Rolnick and Grunewald estimated the annual real internal rate of return at 16 percent, 4 percent for the participant and 12 percent for the public.[24] A more recent comparison of four longitudinal studies of early childhood programs, including the Perry High/Scope study, demonstrated benefit to cost ratios between three and seventeen dollars per dollar invested and internal rates of return from 7 to 20 percent.[25]

The comparison of investment returns is itself a measurement of efficiency, but the Rolnick/Grunewald analysis also used efficiency as an indicator of student compliance. During their K–12 years participants were less likely to be held back, more likely to graduate, and less disruptive in class.[26] Efficiencies in grade advancement and graduation are straightforward: starting with so many students, the more of them who advance each year, or graduate, the higher the efficiency of their progress through school. The limit is clear: more students cannot advance or graduate than were enrolled. Classroom disruptions are a different matter. The claim that they are inefficient is either arbitrary, depending on a comparison of ratios of disruptions to grades or graduation rates, or it is assumed to be inherent in the nature of efficiency itself—that interruptions or disruptions are obstacles to be overcome. Finding efficiency in fewer disruptions illustrates its connection with administrative control. As in *Pickering* and in *Connick*, efficiency is here a measurement of management success.

This finding is borne out in the remarks of conference participants, some of whom were wary of placing too great an emphasis on the cognitive advantages of early childhood education and who noted that its greater impact lay not in changing cognition but in changing behavior. UCLA economist Janet Currie argued that increasing funding for programs such as Head Start, the federal early childhood development program, may improve test scores but does not address behavioral issues. Heckman agreed, arguing that early childhood programs were important because they socialized children to cooperate, not because they improved IQ scores. Advantages in IQ fade in elementary school, as other students catch up, but "social skills and motivation have large payoffs in the job market."[27]

The business connection illustrates the use of efficiency in extending managerial authority into new areas. Conference participants did believe they were on to something new. Rip Rapson, McKnight Foundation president, found it novel to discuss early childhood development in business terms. "When we talk about the well-being of infants and small children, we're not used to thinking of business productivity or state profitability or even public policies; and we're not used to talking about the rates of return or opportunity costs you'll hear about today."[28] The *StarTribune* enthusiastically endorsed greater investment in early childhood education and called for more active business leadership in the effort. The Minnesota State Legislature's early childhood caucus has itself called for more business leadership, and the Minnesota Business Partnership, made up of the CEOs of Minnesota's 105 largest companies, has made improving education its key issue.[29] The Minnesota School Readiness Business Advisory Council, formed in March 2004 and subsequently renamed Minnesota Business for Early Learning, held its first meeting at the McKnight Foundation; representatives from Minnesota's largest and most influential businesses gathered to hear T. D. Churchwell, president of Oklahoma's leading electricity provider, describe their importance to early childhood education. "No group can have greater influence than the business community," he averred.[30]

Many in the local community were pleased to see the issue raised and wrote to praise the *StarTribune*'s coverage of the conference, but their letters did not endorse businesses as leaders in the effort. Paul Gruchow of Duluth took exception to the paper's endorsement of business support for early childhood programs, and especially to its appeal for business leadership, on the grounds that such programs were good for business; such an appeal was "avaricious," he wrote, and he believed it would be better to do what was good for children "simply because it's good, whatever the effect on corporate bottom lines."[31] John Sens of Eden Prairie, a fast-growing suburb southwest of Minneapolis, objected to the extension of professional authority into children's earliest years, even into the womb; parents may not be willing to surrender more "of the best hours of [their] children's lives" to child care experts.[32]

The Minnesota proposal treats early childhood development programs as efficient investments and has extended the efficiency ratio into an area where it had not previously been applied, describing such programs as "a good buy" and a cost-effective strategy for economic development. As Heckman described it, the returns are so good that it is one of the few investments in which there is no trade-off between efficiency and equity.

Efficiency, Flow, and Computing

Dell is the world's largest computer manufacturer, and it continues to operate in the United States at a time when all other major computer makers have sent their manufacturing business overseas. That is only possible, its executives say, because they are fanatics about efficiency in the old style: surveillance and discipline. Dell's story illustrates the pressures that a globalizing economy places on workers, through demands for continually increasing efficiency. Dell's critics see in it a tension between efficiency and innovation and charge that employees who are pressed for minor efficiencies do not have a chance to think in a larger way.

Time and motion studies are alive and well in Dell's enormous assembly plant in Austin, Texas. Workers are videotaped to see how they move and trained by watching videos of workers who are so fast that they can shorten the time it takes to assemble a computer by a few seconds. Those who are not fast enough are fired. Gary Rivlin, business reporter for the *New York Times,* called Dell's Austin plant "dazzlingly efficient" and its managers the "wizards of efficiency." Executives come from all over to see if they can copy what Dell does; the best-selling *New York Times* columnist Thomas L. Friedman sees in Dell the epitome of supply chain management in a global, "flat" economy.[33]

The best example of the search for efficiency through technology in the young twenty-first century is the computer. Dell illustrates the persistence of typically industrial forms of efficiency, such as time and motion studies, alongside new techniques relying on information and telecommunications. Computing and the use of information technologies illustrate the continued application of the concept of efficiency in a variety of ways but most significantly in smoothing out disruptions. The transformative potential of efficiency through computing is celebrated, alongside evidence that it also centralizes managerial control of various processes and allows firms to search for labor markets that offer the least chance of resistance. David F. Noble's study of the development of numerical control technology provides a historical example of information technologies used to create efficiencies by eliminating disruptions; Noble treated these technologies as a form of worker control, designed to maintain the flow of work. He demonstrated that efficiencies achieved through numerical control machinery, designed by control engineers at MIT in the late 1940s, allowed management both to avoid the disruptions that skilled workers could present, in the form of slowing down work, or "pacing," and to standardize work processes and thus the quality of the product.[34] Efficiency, standardization, and quality were defined by the absence of interference.

Other examples of computing and information technologies used to promote efficiency by preventing disruptions come from the work of analysts and management consultants Shoshana Zuboff and James Cortada and sociologist Manuel Castells. Zuboff writes from the perspective of a consultant advising management on strategies for adopting and exploiting information technologies. Cortada analyzes computing's role in a variety of industries. And Castells considers what it means to make a life in an information society.

Efficiencies through computing are the basis of Zuboff's concept of "infomating," created in a study of a number of companies that introduced information technologies in the 1970s and 1980s. Zuboff describes how computing should be implemented if its full efficiencies are to be realized. The concept of infomating describes the ability of information technology to create systems that mimic or replace human physical gestures and, in turn, to generate information about the productive and administrative processes to which it is applied. It covers both the input and output end by turning the process itself into an infomated product.

Two relevant points emerge from Zuboff's study: that workers are left highly visible, in what she calls the "information panopticon," and that workers' reactions to information technologies must be carefully managed if the technologies are fully to succeed. Infomating requires workers highly trained at producing and manipulating a "dynamic, fluid electronic text" as their main product, a text that also leaves them highly visible. Zuboff compared the implementation of computerized oversight systems at two companies, a paper mill named, for the purposes of the study, Cedar Bluff and a telecommunications company, Metro-Tel, also renamed. At Cedar Bluff a system called Overview automatically recorded instrument values across the plant, doing away with the need to rely on employees to record a reading. The technology was managed in an open way so that managers and operators could see the same data, and managers, although regretting some loss of their authority, worked to remain transparent. Cedar Bluff succeeded in fostering a community with a sense of collective responsibility. At MetroTel, by contrast, a Work Force Supervisory System introduced to manage the scheduling of crafts workers, who needed to travel over a large geographic area to maintain and repair electronic switches, was used to "drive" and "push" them, and they were not privy to knowledge about how the system generated assignments. Operators began to look for ways around the system, and management and workers never achieved a sense of joint responsibility and collective awareness. Zuboff describes Cedar Bluff's system as "horizontal" rather than hierarchical; in contrast to MetroTel, its "shared universal transparency" created a sense of mutual responsibility.[35] Cedar Bluff was successful at managing potential disruption.

It is worth noting how Zuboff describes the mechanisms that reinforce cooperation in a transparent workplace. They turn on a sense of visibility. Principal among them is shame, both a fear of being caught unaware and unprepared and a fear of having one's errors exposed; it is a powerful inducement to conformity. Another inducement is the psychological satisfaction provided by a sense of shared goals and teamwork. Whether cooperation is reinforced by one's peers or one's boss, whether hierarchical or horizontal, the point remains: the efficiencies achieved rely on close observation of work in process and on methods designed to prevent interference by carefully managing how new information technologies are deployed.

Flow is an essential characteristic of two of what James Cortada identified, in an influential study, as the most important changes made possible by efficiencies through computing: tighter supply chain management and centralized control of continuous processes. The chain of supply includes all links between materials, where they are first extracted, and the delivery of finished products or services. Managing supply chains is the most widespread use of computing in American manufacturing; it marks not a postindustrial development but an industrial application. Transportation scheduling, in particular, also makes extensive use of the Internet. Computing, combined with telecommunications, makes it possible to integrate activities across vast stretches of space and to coordinate the design, production, and shipping of products; supply chain management is, in essence, the practice of forecasting and scheduling. Hard disk drive production is one of the most competitive fields within the electronics industry. Supply chain management techniques have allowed the top producers to locate production in areas of cheaper labor, far from their research and development operations, and still be able to meet customer timelines.[36] Disruptions within a supply chain can have dire consequences for a firm, especially one that practices just-in-time sourcing and no longer keeps inventory in stock and on hand. The consequences may also be grave for a region whose workers depend on that industry.

Efficiency was also the driving force behind the adoption of information technologies in other cases that Cortada analyzed. In the petroleum transport industry efficiency gains were achieved through the quite literal control of flow, through increasingly centralized management of pipeline systems. Oil is expensive and piped vast distances, and transporting it is a process industry, in which work depends on continuous flow. The less disruption, through leakage, flow fluctuation, or other sorts of interference, the more efficient the system. Computing accompanied increased use of closed-loop systems, relying on pneumatic tools for maintaining regularity of flow and generating data for management by a centralized control facility. Valves and switches, as well as tank level and pipe pressure gauges, are read and controlled by the central facility, which can dispatch labor or equipment as needed.[37] The petroleum transport industry example underscores the continuing role of close observation in efficiency measures. Surveillance is not directed primarily at workers but at the system, although the observation techniques do reveal information about the activities of those who work within its reach.

The importance of eliminating disruption is apparent in two examples from the work of sociologist Manuel Castells. These cases also highlight distinctions between efficiencies achieved by workers and efficiencies achieved by management, or between efficiencies of balance and stability and efficiencies of growth. The first is the management system of the automaker Toyota, called "Toyotism," much celebrated in the 1990s for its efficiency. Central was management's assumption of "'five zeros': zero defect in the parts; zero mischief in the machines; zero inventory; zero delay; zero paperwork." As Castells points out, the system also required an inflexible approach to workers: disruptions could not be tolerated because the system was so fine-tuned that it left no room for them— meaning, as Castells puts it, that the system was premised on "total control over labor." He concluded that Toyota's was "a management system designed to reduce uncertainty."[38]

Castells also argues that information technologies have created a new and fundamental distinction among workers, between "a core labor force . . . and a disposable labor force." New technologies increase efficiency by providing flexibility and adaptability to the owners and managers of capital, allowing them to circumvent obstructions such as those represented by the labor unions of Europe and the earlier United States. One result is that working life has become highly vulnerable, even for knowledge workers, who are subject to a "virtual enterprise" that can now "assemble and disperse labor anywhere, anytime."[39] Core workers perform symbolic manipulations, using information technology; disposable workers are those who are subject to hiring and firing, or outsourcing, on the basis of demand and labor costs in their local labor markets. Workers, unless performing precisely as required, themselves become obstacles to be overcome.

These examples illustrate the significant tension that exists between the structures and constraints that underlie the globalized world and the freedoms it promises. The tension is particularly clear in the cases from American law, in which freedom of speech was weighed against the administrative pursuit of planned goals. In Rosenbaum's critique the tension arose between the freedom to tell a story fully and the administrative imperative to process legal cases. The tension is less apparent in the business movement for early childhood education, in which it lies between poverty and successful integration into a functioning economic system. Poverty does not seem to be a freedom but is, rather, a constraint, until it is recognized as the point of origin for the many sorts of nonconforming behavior cited in the studies of the social returns of early childhood education. Efficiency lies in governing such nonconforming behavior. Two kinds of freedom were at issue in the examples taken from computing: the free movement of capital and materials, supported by physical and legal structures, and the freedom or self-determination of workers, who could resist or facilitate central productive activities. Castells showed how workers have become constrained by the flexibility of the new labor market, which renders them vulnerable; Zuboff described management techniques intended to constrain resistance to the adoption of information technologies. These examples suggest that efficiency remains crucial in a global economy and not only because traditional manufacturing operations continue. It remains essential as a tool of avoiding disruption, allied with administrative and material structures that remain susceptible to interference.

Conclusion

The Future of Efficiency

Although Sadi Carnot was, by all accounts, a melancholy and solitary young man, he did manage to end up among the throngs of mechanics under police surveillance in Paris's radical, working-class third arrondissement, at the program of public lectures inaugurated in 1819 at the Conservatoire des Arts et Métiers, during Gérard-Joseph Christian's tenure as director. The legacy of revolution was still fresh, the heir presumptive to the throne recently assassinated, and the Parisian mechanics whose workshops lined the streets of the quarter had proven themselves able opponents of monarchy and privilege. Among the lecturers who delighted their listeners with jabs at the government was the industrial chemist Nicolas Clément; otherwise, he was apparently a lackluster speaker, but he became a friend of Carnot and a great influence. In their lifetimes Clément's fame far outstripped Carnot's—Carnot had none, really; those who knew the Carnot name knew his father, Lazare Carnot, engineer, machine theorist, politician—but now, nearly two hundred years later, the name Carnot means Sadi, and he is a hero of physics and engineering textbooks. His name also means a cycle, the Carnot-cycle, of perfect, ideal efficiency.

Carnot's contribution was a small book published in 1824 on the motive power of heat; within a generation it was judged a work of genius, but that was too late for the author, who had died of cholera in 1832, still a young man. Carnot did not himself characterize the cycle as efficient; later researchers did that, as they worked out the fundamental laws of thermodynamics to which Carnot's work pointed. So obscure had Carnot's work become that William Thomson (Lord Kelvin) had difficulty finding a copy once he read mention of it; the Parisian booksellers knew only of Lazare Carnot's work.[1]

What Carnot wanted was a theory that could account for the performance of heat engines as completely as theory accounted for machines driven by muscle power, or water or wind. Mechanical theory allowed such machines to be analyzed "down to their last detail," he wrote. "Every event is predictable, all possible movements are in accordance with established general principles which are applicable in all circumstances."[2] This remained theory, it must be stressed, and concerned machines only in the ideal or abstract case; Christian himself protested that too much of machine theory was practically useless, echoing the sentiments the century before of British engineer John Smeaton, who had blamed theory for misleading practical men who had to work with real materials. Carnot, nonetheless, had described the ideal basis for efficiency: a process fully understood and fully predictable, disrupted neither by unknown movements nor by particular or unique ones. He had described a system that was susceptible to planning and control.

Beneath each example of efficiency in this book lies the attempt to control a changing situation, by bringing it into conformity with a vision of how the world works. In most cases the attempt combined two allied but distinct senses of efficiency, one static and emphasizing stability and balance, the other dynamic, linked with growth and transformation. The distinction underscores how deeply efficiency was embedded in modernity and suggests ways in which, modern though it is, it will endure even in a world hailed as modernity's successor.

Control is as slippery a term as *efficiency*, and there are many ways to think about it. Controlled experiments, statistical controls, feedback and control—each of them come with fully articulated theories. Then there are loose, colloquial meanings: feeling out of control, losing control of the car, avoiding a controlling person. In common is the sense of control as regulating or governing something, making it conform to expectations. What makes efficiency modern is not only its association with control, for the use of rational techniques and a reliance on planning also characterize the modern. What efficiency added was a measurable way to assess control. It was managerial, governing both processes and their outcomes; its aim was to regulate behavior, natural, human, or machine.[3] This sort of close control was associated not only with Taylor's efficiency efforts but also with the Fordism of the assembly line and the intrusions of time and motion researchers. Observing and ordering were not themselves the primary objectives; they were instead part of larger visions of how the world should be ordered. A single concept, efficiency, brought together both transformative visions and the specific methods to achieve them; it united hopes of dynamic change with the often restrictive measures that might bring it about.

The expectations behind particular uses of efficiency are thus crucial in assessing its character. A number of visions were in play in the instances analyzed here: some were normative, portraying obligations or duties; others descriptive, about how the physical world could be expected to behave. A vision of the ideal roles of people and machines in society characterized the normative models; descriptive models took the form of mechanical or scientific theory. In some instances the operating vision was of motion itself—what caused it and what should move and how much. The control of water flow was essential to eighteenth-century trials of waterwheel performance, as was the control of vibration or shaking in the new and heavy textile machines that symbolized industrial progress. Behind both was a view of motion or motive force as something to be conserved. In human terms the vision was of the necessary roles of managers and employees and the need to achieve stability among the workforce and in work production. The Gantt chart expressed its author's belief that laborers in the Progressive era United States needed to be taught how to work with diligence and application, just as motion researchers in Weimar Germany believed workers could not move their own bodies efficiently and thus needed props and restraints. Among the descriptive and theoretical constructions were mechanical models of matter in motion—and, later, conceptions of the interrelationship between motion, heat, and energy, as expressed in the nineteenth-century science of thermodynamics—and the equally foundational theory of Charles Darwin, which accounted for the enormous diversity of living beings to be encountered on the earth. In sum, according to these visions, water should return to the height from which it falls, motive power or force should not be wasted, machine theories should address mechanical practice, workers should do as they are told, energy should not be wasted, struggle should yield better-adapted forms of life, people should control themselves, and inhumane practices should be inefficient.

A distinction between rhetoric and reality, or between goals or visions and available tools, is not unique to efficiency. What was unique was how it combined both elements in a single measure. It combined the vision, as output, with an assessment of available tools, as input. Efficiency thus facilitated a particularly rigorous form of planning. Planning was a central feature of modernity and modern technological society, and the developing concern for efficiency best embodied the idea in its technological form: the surveying and apportioning of resources to achieve a desired result. Technology, itself closely linked with modernity, is foremost an expression of the human desire to have effect and to influence surroundings and events. Efficiency offered more than a way to compare output to input, as a way of assessing effectiveness; such comparisons had been made long before efficiency was named and codified in theory. What efficiency did when it became a general concept, rather than an incidental tool, was to make fundamental the assessment of almost any action or process on the basis of the same units and qualities it had started with and nothing else. It created a closed intellectual system. The amounts of what went in and came out would differ, but the fundamental categories of measurement, the units, would not. The focus would not shift.

The notion of a process fully understood and fully predictable was quintessentially modern. It expressed a faith in the powers of the human intellect to comprehend the world and the belief that the world itself did indeed correspond to the categories and methods of human comprehension. Carnot illustrates the matter most strongly; the faith that the world may be fully apprehended by the human mind, if it follows the right epistemological steps, lies behind the move from understanding a process to predicting it. Things may be understood in a variety of ways—an understanding might bring intellectual satisfaction or religious or philosophical comfort—but to move beyond comprehension, to go beyond the mental construction of what something means to predicting what it will do, required a robust sense of the material potential of human rationality. It required the belief that one could intervene with success.

This idea was not straightforward; people's inability to bring outside circumstances into line with their own ways of thinking has long been a source of frustration. Deliberate and purposeful causal agency can be difficult to achieve. Material conditions, in addition to human intentions, act to influence outcomes; such conditions may also defy positive, rational understanding, meaning that outcome is not governed by how material processes and conditions are intellectually conceived. The challenge of bringing an intellectual understanding successfully to bear on the material world can be seen in the often simple measures prescribed for efficiency. Christian, despite a sophisticated understanding of mechanisms in motion, still found it necessary to emphasize something so basic as bolting a machine to the floor. Thoroughly masticating one's food and briskly toweling after a bath were among Purinton's prescriptions; they supported his conception of personal efficiency through self-mastery but, in comparison to his goals, seem almost banal.

Prediction makes it possible to plan for successful interventions, the sort of planning that was at the core of efficiency. It allows the conditions of a process to be manipulated to increase the likelihood that it will achieve a desired outcome; the tailrace of a waterwheel is designed to minimize turbulence in the water, the moving parts of a spinning machine carefully integrated to operate smoothly and without jarring impact, the workers in a facility undergoing automation given incentives to accept a new system. Planning was the motive for both Smeaton and the Franklin Institute researchers of Philadelphia in their celebrated studies of waterwheel performance; they had in sight the practical decisions engineers and mechanics faced—where to put an engine, what size it should be, at what speed it should run. Carnot himself envisioned a time when heat engines might transform industrial technique; he had in mind the steam engine but freely admitted the potential for new machines, of which he could scarcely conceive. Only with a full understanding of how heat works would people be able to design engines that took full advantage of the powers available in the fuels they used;[4] only then would they be able adequately to plan.

Efficiency was not primarily about knowledge; it was a method by which intellectual constructions could be given material force. In its dimensionless and comparative form, as a percentage without units, efficiency applied equally to rational and mechanical constructions and allowed them to be analyzed and measured on commensurate and parallel scales. Engineers and mechanics sought to increase their effective mastery of the material world by constructing machines whose performance not only paralleled the ways it was intellectually understood but also, and simultaneously, minimized the influence of material behaviors for which no intellectual parallel existed. The uniformity and predictability of static efficiency relied on the elimination of distractions or diversions that were too complex to admit of measurement or management; inherent in the concept of efficiency was the goal of eliminating or minimizing resistance to methods of rational control.

Two among the examples in this book concerned efficiency as a force that was unplanned and undirected: Darwin's metaphor of the efficiency of natural selection and Fogel's admission that the efficiency of slavery in the antebellum United States resulted from the brutal practices of gang labor. It is significant that both of these examples are ones of wounding—the driving of the wedge between species that survive and those that do not and the whip that forces intense, dehumanizing labor. Not only does its variety make efficiency difficult to characterize historically but also its duality, especially the striking contrast between celebrations of efficiency and cries against the deep hurt it has caused. Fogel experienced both, having gone from viewing efficiency as the necessary byproduct of humane treatment and mature self-discipline to seeing within it real practices of brutality.

Alongside the duality of efficiency's static and dynamic forms is thus a deeply mixed reckoning of its moral value. This is neither a paradox nor mere ambivalence; the polarity between celebrating efficiency and decrying it as exploitative illustrates how profound are the meanings people have attached to it. Fogel went from one extreme to the other. He expected to find American slavery inefficient, as would befit a tyrannous system, at least in the way he saw the world, and he experienced a crisis when his data turned out to prove the opposite. He resorted to reevaluating slavery itself, accounting for its efficiency through the good work ethic of the slaves themselves and the relative absence of physical coercion. But as the evidence mounted that slavery's productivity came not from good management and willing workers, who were well treated and well fed, but from gang labor so intense that it required brutal enforcement, Fogel had again to reevaluate his commitment. He finally, and painfully, sundered efficiency from his ideas of what was moral and good. Fogel had believed efficiency was something to celebrate, but he came to see in it a powerful tool of repression.

Characterizing efficiency requires taking seriously both the benefits and the deep hurts it has offered. Its endless variety tempts historians to treat efficiency piecemeal, as a product of particular contexts at particular times, but distinguishing between its different forms and the values it has accrued makes it possible to move beyond considerations of local context to examine the broader function of efficiency within modern industrial culture. Measurements of efficiency have been key in the development of machines that deliver more power while using fewer resources and in increasing the mobility and comfort of untold numbers of people throughout the world.

The sticking point is efficiency and barbarism, specifically the efficiency and inhumanity of Nazi Germany's murderous practices during World War II.[5] So what of efficiency and the Holocaust? The apparent answer is that it was a tool, an instrument used by Nazi Germany in pursuit of its goals, just as tools are used by everyone, however horrifying or inhumane their aims. It was, apparently, merely a means to an end. But looking closely at what happened suggests a more powerful role for efficiency, both in the act of genocide and in the analysis of it. Murderous violence against Jews and other enemies of the reich preceded the measures historians describe as efficient, and nonindustrial ways of killing continued throughout the regime. Much of the killing was brutal, messy, and expedient, not industrial.

The factories of death in the concentration camps, if we think of them in terms of efficiency, demonstrate the centrality of racial ideology to the Nazi regime. One may speak of pockets in which efficiency was calculated and is recognizable not in a loose sense but as a rigorous technique. Crematoria II and III at Birkenau, in the Auschwitz complex, are examples: they were the only fully industrialized killing systems, arranged for the continuous flow of victims and

their bodies, from undressing chambers to the removal of their ashes. Nevertheless, industrial methods do not always imply efficiency; assembly lines may be
used, and processes organized, without efficiency being calculated. But the industrialized methods do betray the regime's commitment to the most extreme
form of racial violence, a commitment revealed even more strongly if the lens of
efficiency is drawn back, and if the forms of measures of output and input are
widened, to include not only specific techniques or practices but also the goals
and resources of the regime, or its desired output and available inputs. The
desired output was a Germany free of Jews; the input came to be the most pre-
cious resources of the reich. The regime continued to put material resources into
murder, even when they were urgently needed elsewhere. Even late in the war,
for example, crucial supply lines continued to carry trains transporting victims to
the murder camps, instead of basic supplies and war materiel that were desper-
ately needed at the front. Dan Diner describes it well, the "endless rolling freight
trains loaded with human victims, silhouetted against a backdrop of German military chaos."[6]

The techniques associated with efficiency, as embodied in the methods of
industrial murder, were undertaken when the final solution had become plain,
and it was made clear that the regime's purpose was to annihilate its internal enemies. Despite all the horrible mistreatment, not until sometime in 1941 was the
decision to kill all Jews actually undertaken. Efficiency allows that purpose to be
identified. The regime put its dwindling resources into the killing of powerless
victims at home. The efficiency of the Holocaust was not merely instrumental; it
demonstrates how central to Nazism was its racial ideology. The regime was
premised on purifying the Aryan race and its living spaces; thus, it also required
the removal of peoples described as degenerate or polluting. The removal was not
merely a consequence of the quest for utopia; it constituted the very vision of
utopia itself. Consistent within the Nazi movement, and within the regime, was
the continued assault—verbal, rhetorical, and increasingly physical—on the
groups against which Nazism defined itself: social, physical, and criminal so-
called degenerates; Roma and Sinti, the unsettled wanderers; and Jews. Behind
the genocide that was the Holocaust was the powerful ideal of its perpetrators:
that they were building a new society and that achieving it demanded the actions
they took. But the Holocaust was not merely a necessary measure once one
became committed to the vision; it was the ultimate expression of the vision itself. Central to Nazism was the taking of control to create a new society, and efficiency, insofar as it is an instrument, is powerfully allied with control and the
achieving of visions.

+ + +

Langdon Winner argued that discussing technology in terms of efficiency, along with other technical criteria such as productivity and modernization, is to miss "a decisive element in the story."[7] Technical criteria, Winner suggested, do not account for technologies and technological systems that may not only be used to gain and perpetuate power but which may embody claims to power and authority in their very design. Efficiency, however, even in its technical form, was a tool of control and not a mere technical measurement, disciplinary at its inception and increasingly political after it had reached conceptual maturity. As a measurement, it has an apparently objective form, but its history is as a tool designed to make the natural and human worlds conform to the way in which they are intellectually understood.

Efficiency through technological means continues to be important in the twenty-first century. Control of motion and surveillance techniques remain influential, and the duality between efficiency's static and dynamic forms may be observed even in the use of information technologies celebrated for their flexibility and lack of hierarchy. Historically, efficiency served as a tool to minimize turbulence, and it will likely continue to play the same role as long as people continue to plan.

The need for planning is precisely what some theorists of postmodern culture see as increasingly passé, arguing that efficiency will be superseded as society is reorganized in new ways. An emerging networking culture, new forms of cooperative management, and the reliance on flexible production and the free flow of capital will mark the end of hierarchically managed and directed systems of efficiency. Such hierarchical control was famously associated with Frederick Winslow Taylor's system of scientific management and the European concept of rationalization;[8] whereas in these systems there was "one best way" to do something efficiently—in a phrase associated with Taylor—now there are many ways to do it, and without central control or surveillance.

Centralized control and surveillance are indeed crucial components of modern technology, and they have come under sustained criticism, as has the concept of modernity itself, since World War II. Whether modernity may adequately be described as a project of emancipation or as a period of progress through human reason is in serious doubt. Especially criticized is the inability of modern narratives of progress to give full and serious weight to cultures and experiences that lie outside its main story line, of which progress through increased efficiency is one. Yet the modern emphasis on human reason remains potent, as does a focus on humanity's ability to manage or create its own environments, its "coming of age," in other words. Stephen Toulmin, in an extension of modern planning, has suggested ways to humanize and preserve modernity by bringing into it more diverse forms of human experience and community.[9] Romano Guardini, writing in the wake of World War II, argued that technology will always require modern planning: in modern technology people see a world that "does not run itself" but which must be run, a world that demands a ruler.[10] As long as planning remains central, so will efficiency.

Central to efficiency has been a distinction between two allied but quite different meanings, woven together in an apparent paradox. Efficiency described both performance within strict limits and performance that exceeded limits; it was simultaneously a measure of conservation and of growth. Efficiency was a matter of turning energy into useful work, for example in the natural gas engine installed early in the last century at the Pennsylvania State Hospital, where there was no question of its efficiency reaching 100 percent—the laws of thermodynamics described its limits—and it certainly could not achieve anything greater. In the same years efficiency became part of a transformative rhetoric of changing the world, a use echoed in Roosevelt's New Deal and objected to by Hoover. Efficiency described both conservation and transformation. On one hand, it was an accounting principle, comparing what went into a machine or process with what came out and measuring how much of the original input had been conserved. On the other hand, efficiency was creative and dynamic, yielding growth through careful management and bringing as its reward not merely conservation but increase. These two meanings were woven together in a variety of uses stressing both the conservative character of efficiency and its dynamic and creative potential.

Consistent throughout its history has been the equation of efficiency with mastery, from the Aristotelian system of causes through medieval conceptions of the nature of the Christian God. In modern terms efficiency also offered human mastery, through techniques of surveillance and discipline and through accounting principles to track the use and transformation of materials. Efficiency also helped in mastering other people. After the mid-nineteenth century it meant something different for managers than for those who were under management's control. Managers had access to the rewards of efficiency as a tool of growth, unbounded by the rational and zero-sum calculus of a mere ratio of output to input. Those who were managed remained subject to that ratio, providing the stable and balanced ingredients that managers manipulated in pursuit of growth. Efficiency meant mastery; it also meant being mastered.

Introduction: The Varieties of Efficiency

1. Joseph Conrad, *Heart of Darkness*, in *Great Books*, ser. 1 (Chicago: Great Books Foundation, 1985), 1:155; David Landes, *The Unbound Prometheus* (Cambridge: Cambridge University Press, 1969), 26.

2. Jacques Ellul, *The Technological Society* (New York: Knopf, 1964), 20–21, 74, 427; Lewis Mumford, *Technics and Civilization* (1934; rpt., New York: Harcourt, Brace and World, 1963), 387, 390.

3. *Oxford English Dictionary*, "Efficiency-Date Chart," http://dictionary.oed.com/cgi/entry_main/50072381?single=1&query_type=word&queryword=efficiency&first=1&max_to_show=10.

4. See Anson Rabinbach, *The Human Motor: Energy, Fatigue, and the Rise of Modernity* (New York: Basic Books, 1990); Mary Nolan, *Visions of Modernity: American Business and the Modernization of Germany* (New York: Oxford University Press, 1994); G. R. Searle, *The Quest for National Efficiency: A Study in British Politics and Political Thought, 1899–1914* (London and Atlantic Highlands, NJ: Ashfield Press, 1990); and Sheila Faith Weiss, *Race Hygiene and National Efficiency: The Eugenics of Wilhelm Schallmayer* (Berkeley: University of California Press, 1987).

5. Searle, *Quest for National Efficiency*.

6. See Eric D. Weitz, *A Century of Genocide: Utopias of Race and Nation* (Princeton: Princeton University Press, 2003), 143. This is an example frequently voiced by my students in class. *Wikipedia* devoted an entire section to efficiency in its entry on the Holocaust (http://en.wikipedia.org/wiki/Holocaust#Efficiency, July 3, 2006).

7. Michael Thad Allen demonstrates that of all the apparatus of extermination in the Holocaust, only two crematoria in Birkenau may accurately be described as "modern industrial factories"; see "Modernity, the Holocaust, and Machines without History," in Michael Thad Allen and Gabrielle Hecht, eds., *Technologies of Power: Essays in Honor of Thomas Parke Hughes and Agatha Chipley Hughes* (Cambridge, MA: MIT Press, 2001), 198; Michael Thad Allen, *The Business of Genocide: The SS, Slave Labor, and the Concentration Camps* (Chapel Hill: University of North Carolina Press, 2002); Christopher R. Browning,

Ordinary Men: Reserve Police Battalion 101 and the Final Solution in Poland (New York: HarperCollins, 1992).

8. *New York Times,* November 1, 1932, 1, 12–25; Herbert Hoover, "Campaign Speech at Madison Square Garden, New York City, October 31, 1932," *The State Papers and Other Public Writings of Herbert Hoover,* 2 vols. (New York: Doubleday, 1934), 2:408–28.

9. Soma Golden, "New Slavery Book Kindles a Dispute," *New York Times,* May 2, 1974, 49, 57; Robert Reinhold, "100 Scholars Gather to Debate Two Economists," *New York Times,* October 28, 1974, 61.

10. Martin Carcason, "Herbert Hoover and the Presidential Campaign of 1932: The Failure of Apologia," *Presidential Studies Quarterly* 28 (1998): 349–66.

11. Robert William Fogel and Stanley L. Engerman, *Time on the Cross: The Economics of American Negro Slavery* (New York: Norton, 1974); see also Fogel and Engerman, "Afterword 1989," in 1989 edition of *Time on the Cross* (New York: Little, Brown, 1989), 265–75; and Robert William Fogel, *Without Consent or Contract: The Rise and Fall of American Slavery* (New York: Norton, 1989).

12. See, for example, Cecelia Tichi, *Shifting Gears: Technology, Literature, Culture in Modernist America* (Chapel Hill: University of North Carolina Press, 1987), 63; Martha Banta, *Taylored Lives: Narrative Productions in the Age of Taylor, Veblen, and Ford* (Chicago: University of Chicago Press, 1993); and John M. Jordan, *Machine-Age Ideology: Social Engineering and American Liberalism, 1911–1939* (Chapel Hill: University of North Carolina Press, 1994), 37, 44.

13. Theodore M. Porter, *Karl Pearson: The Scientific Life in a Statistical Age* (Princeton: Princeton University Press, 2004), 97, 220.

14. This is a central argument in two recent collections addressing historical issues in modernity and technology: Thomas J. Misa, Philip Brey, and Andrew Feenberg, eds., *Modernity and Technology* (Cambridge, MA: MIT Press, 2003); and Michael Thad Allen and Gabrielle Hecht, eds., *Technologies of Power: Essays in Honor of Thomas Parke Hughes and Agatha Chipley Hughes* (Cambridge, MA: MIT Press, 2001). Misa, for example, finds Marx's analysis to be industrially well informed but criticizes Weber, Horkheimer, and Habermas for dealing with technology only in the abstract.

15. Alain Touraine, *Critique de la modernité* (Paris: Fayard, 1991).

16. Michael Adas, *Machines as the Measure of Men: Science, Technology, and Ideologies of Western Dominance* (Ithaca, NY: Cornell University Press, 1989), 402–18.

17. Manuel A. Vásquez, *The Brazilian Popular Church and the Crisis of Modernity* (Cambridge: Cambridge University Press, 1998), 23; Immanuel Kant, "An Answer to the Question: What Is Enlightenment?" in *Postmodernism: A Reader,* ed. P. Waugh (London: Edward Arnold, 1992).

18. Carl Mitcham, *Thinking through Technology: The Path between Engineering and Philosophy* (Chicago: University of Chicago Press, 1994), 225–28.

19. Donald Cardwell, *Norton History of Technology* (New York: Norton, 1995), 83–91; Robert K. Merton, "Science, Technology, and Society in Seventeenth Century England," *Osiris* (1938): xx–xxi, 521.

20. The other three Aristotelian causes dealt with the material and formal aspects of change, and with its ultimate purpose, or final cause.

21. T. Spencer, *Logick* (1628), 31.

22. Edward Grant, *Physical Science in the Middle Ages* (Cambridge: Cambridge University Press, 1977), 31; Marilyn McCord Adams, *William Ockham* (Notre Dame: University of Notre Dame Press, 1987); Frederick Charles Copleston, *A History of Medieval Philosophy* (New York: Harper and Row, 1972); Étienne Gilson, *Saint Thomas Aquinas and Philosophy* (West Hartford, CT: Saint Joseph College, 1961).

23. Donald Worster, *Nature's Economy: A History of Ecological Ideas*, 2nd ed. (Cambridge: Cambridge University Press, 1994), 37; Ernesto Screpanti and Stefano Zamagni, *An Outline of the History of Economic Thought* (Oxford: Clarendon Press, 1993), 20–23.

24. See Richard Rorty, *Philosophy and the Mirror of Nature* (Oxford: Blackwell, 1980); Philip Brey, "Theorizing Modernity and Technology," 58–62; and Arie Rip, "Modernity and Technology—an Afterword," 363–65, both in Misa et al., eds., *Modernity and Technology*.

25. Max Horkheimer and Theodor W. Adorno, *Dialectic of Enlightenment: Philosophical Fragments*, ed. Gunzelin Schmid Noerr, trans. Edmund Jephcott (Stanford, CA: Stanford University Press, 2002), throughout, quote on 67–68. Jürgen Habermas has remained a steadfast critic of this conception of reason; see Habermas, *The Philosophical Discourse of Modernity* (Cambridge, MA: MIT Press, 1987), and *The Theory of Communicative Action*, vol. 1: *Reason and the Rationalization of Society* (Boston: Beacon Press, 1984).

26. David Harvey, *The Condition of Postmodernity: An Inquiry into the Origins of Cultural Change* (Cambridge, MA: Blackwell, 1990), 10–38; Stephen Kern, *The Culture of Time and Space, 1880–1918* (Cambridge: Harvard University Press, 1983).

27. Franz Reuleaux, *Kinematics of Machinery: Outlines of a Theory of Machines* (London: Macmillan, 1876), 35, 87.

28. Samuel P. Hays, *Conservation and the Gospel of Efficiency: The Progressive Conservation Movement, 1890–1920* (Cambridge: Harvard University Press, 1959); Samuel Haber, *Efficiency and Uplift: Scientific Management in the Progressive Era* (Chicago: University of Chicago Press, 1964); Robert Kanigel, *The One Best Way: Frederick Winslow Taylor and the Enigma of Efficiency* (New York: Viking, 1997); Monte Calvert, *The Mechanical Engineer in America, 1830–1910: Professional Cultures in Conflict* (Baltimore: Johns Hopkins Press, 1967), 275–76.

29. Max Weber, *The Protestant Ethic and the Spirit of Capitalism* (1904–5; rpt., London: Routledge, 2001), 67–69, 107–8.

30. Michel Foucault, *Discipline and Punish: The Birth of the Prison* (New York: Random House, 1977), 136; Philip S. Gorski, *The Disciplinary Revolution: Calvinism and the Rise of the State in Early Modern Europe* (Chicago: University of Chicago Press, 2003), 1–38, 164–72.

31. Weber, *Protestant Ethic*, xxxii–xxxiii, xxxv, 71, 123; Romano Guardini, *The End of the Modern World: A Search for Orientation* (Wilmington, DE: ISI Books, 1998), 167; Ellul, *Technological Society*, 173–83.

32. Norton Wise and Crosbie Smith, "Work and Waste: Political Economy and Natural Philosophy in Nineteenth Century Britain," pts. 1–3, *History of Science*, 27–28 (1989–90): 263–301, 391–449, 221–91; Ben Marsden, "Engineering Science in Glasgow: Economy, Efficiency and Measurement as Prime Movers in the Differentiation of an Academic Discipline," *British Journal for the History of Science* 25 (1992): 319–46.

One · *Static and Dynamic Efficiency*

1. Robert Mallet, "Mr. Mallet's Paper on Water-Wheels," *Journal of the Franklin Institute* 37 (1844): 169.

2. James Whitelaw, "Observations on Mr. Ellwood Morris' Remarks on Water-Wheels"; rpt., *Journal of the Franklin Institute* 38 (1844): 73–74.

3. M., *Journal of the Franklin Institute* 37 (1844): 169.

4. Terry Reynolds discusses various waterwheel trials in the eighteenth and nineteenth centuries as measurements of efficiency, although noting that the term was not in use (*Stronger Than a Hundred Men: A History of the Vertical Water Wheel* [Baltimore: Johns Hopkins University Press, 1983], esp. chap. 4). See also Albert E. Musson and Eric Robinson, *Science and Technology in the Industrial Revolution* (Manchester: Manchester University Press, 1969), esp. chaps. 1–2.

5. Larry Stewart, *The Rise of Public Science: Rhetoric, Technology, and Natural Philosophy in Newtonian Britain, 1660–1750* (Cambridge: Cambridge University Press, 1992), 385.

6. John Smeaton, "An Experimental Enquiry concerning the Natural Powers of Water and Wind to Turn Mills, and Other Machines, Depending on a Circular Motion," *Philosophical Transactions of the Royal Society* 51 (1759): 100–174.

7. Royal Philosophical Society, London, Royal Society Council Minutes, 1747–63, 4:221–22, November 22, 1759.

8. British Library manuscripts, no. 30094, John Smeaton letters to Benjamin Wilson, 1744–64, September 24, 1946.

9. M. Norton Wise, "Mediating Machines," *Science in Context* 2 (1988): 77–114; and Yehuda Elkana, *The Discovery of the Conservation of Energy* (Cambridge: Harvard University Press, 1974), 16.

10. Reynolds, *Stronger Than a Hundred Men*, 241–42, 247–48.

11. Others included the French mathematicians and engineers Antoine de Parcieux and Charles Bossut as well as Christopher Polhem, Swedish specialist in mining engineering (Reynolds, *Stronger Than a Hundred Men*, 218–33).

12. Ibid., 204–18; Norman Smith, "Scientific Work," in *John Smeaton, FRS*, ed. A. W. Skempton (London: Thomas Telford, 1981), 35–57; Colin MacLaurin, *A Treatise of Fluxions* (Edinburgh: Ruddimans, 1742), 728.

13. Smith, "Scientific Work," 46. This was the *vis viva* controversy; see Mary Terrall, "Vis Viva Revisited," *History of Science* 42 (2004): 189–209.

14. "Force," col. 2, in Ephraim Chambers, *Cyclopedia: A Universal Dictionary of the Arts and Sciences*, vol. 2 (London: W. Strahan, 1779).

15. John Smeaton, "New Fundamental Experiments upon the Collision of Bodies," *Philosophical Transactions of the Royal Society* 72 (1781): 337–54.

16. Ibid. Smeaton discussed the laws of motion of perfectly hard, nonelastic bodies as laid down by two "species of philosophers . . . [one] alleging, that there are no such bodies to be found in nature whereon to try the experiment . . . [and the other laying] it down as a rule attending all non-elastic bodies, whether hard or soft, that the velocity of the stroke will be the same in both, *viz. one half* of the velocity of the original striking body" (339–40).

17. Smeaton's famous paper on waterwheels also included analysis of windmills, which, unlike his waterwheels work, produced no fundamental insights or general rules. See Smith, "Scientific Work." Smeaton's two additional papers on mechanics were "An Experimental Examination of the Quantity and Proportion of Mechanic Power Necessary to Be Employed in Giving Different Degrees of Velocity to Heavy Bodies from a State of Rest," *Philosophical Transactions of the Royal Society* 66 (1779): 450–75; and Smeaton, "New Fundamental Experiments."

18. Quoted in Trevor Turner and A. W. Skempton, "John Smeaton," in Skempton, *John Smeaton, FRS*, 11.

19. Ibid.; and Denis Smith, "Professional Practice," also in Skempton, *John Smeaton, FRS*, 217–27.

20. Smeaton, "Experimental Enquiry," 124–25.

21. Smeaton, "Experimental Examination," 458.

22. Smeaton used a variation of Torricelli's and Varignon's laws of fluid flow, Torricelli having established by analogy to Galileo's law of falling bodies that the velocity of liquid flowing through an aperture is proportional to the square root of the height of the liquid. From a velocity of 8.96 ft./sec. Smeaton deduced a virtual head of 15 inches; he used the equation $h = v^2/2g$, a variation of Varignon's mathematical statement of fluid velocity under a static head. Rene Dugas, *A History of Mechanics* (Neuchatel: Editions du Griffon, 1955), 147, 226; see also John K. Vennard and Robert L. Street, *Elementary Fluid Mechanics,* 6th ed. (New York: John Wiley and Sons, 1982), 117.

23. Simon Schaffer, "Machine Philosophy: Demonstration Devices in Georgian Mechanics," *Osiris* 9 (1994): 168–70. See also Richard Olson, *Science Deified and Science Defied: The Historical Significance of Science in Western Culture,* vol. 2: *From the Early Modern Age through the Early Romantic Era, ca. 1640 to ca. 1820* (Berkeley: University of California Press, 1990), 31.

24. Schaffer, "Machine Philosophy," 168–70.

25. Smeaton addressed these inquiries in his two subsequent reports on fundamental mechanical questions; see n. 17.

26. Stewart, *Rise of Public Science.*

27. Clifford Truesdell argued that through the mid-eighteenth century rational mechanics proceeded mathematically, while practical mechanics applied the manual arts. See Truesdell, "A Program toward Rediscovering the Rational Mechanics of the Age of Reason," *Archive for History of Exact Sciences* 1 (1960): 13.

28. John Farey, *A Treatise on the Steam Engine, Historical, Practical, Descriptive* (London: Longman, Rees, Orme, Brown, and Green, 1827), 296.

29. William Waring, "Observations on the Theory of Water Mills, Etc.," *Transactions of the American Philosophical Society* 3 (1793): 144–49.

30. Ibid., 319–21.

31. Wilson L. Scott, *The Conflict between Atomism and Conservation Theory, 1644–1860* (London: MacDonald & Co., 1970), 139.

32. *Journal of the Franklin Institute* 4 (1829): 143.

33. Ibid., 286–87.

34. Smeaton, "Experimental Enquiry," reprinted in *Miscellaneous Papers of John Smeaton, Civil Engineer, FRS* (London: Longman, Hurst, Rees, Orme, and Brown, 1814), 42.

35. R., "Remarks on Friction and on the Power of Heavy Bodies in Motion," *Journal of the Franklin Institute* 4 (1929): 138–43.

36. Reynolds, *Stronger Than a Hundred Men*, 247–48.

37. Eda Kranakis, "Social Determinants of Engineering Practice: A Comparative View of France and America in the Nineteenth Century," *Social Studies of Science* 19 (1989): 5–70.

38. Edwin T. Layton Jr., "Newton Confronts the American Millwrights, or, Action and Reaction Are Not Always Equal," in *Beyond History of Science: Essays in Honor of Robert E. Schofield*, ed. Elizabeth Garber (Bethlehem: Lehigh University Press, 1990), 189.

39. Bruce Sinclair, *Philadelphia's Philosopher Mechanics: A History of the Franklin Institute, 1824–1865* (Baltimore: Johns Hopkins University Press, 1974), 1–19; Adrian Michal McMahon and Stephanie A. Morris, *Technology in Industrial America: The Committee on Science and the Arts of the Franklin Institute, 1824–1900* (Wilmington, DE: Scholarly Resources, 1977), xv.

40. Carolyn Cooper, "A Patent Transformation: Woodworking Mechanization in Philadelphia, 1830–1856," in *Early American Technology: Doing Things from the Colonial Era to 1850*, ed. Judith A. McGaw (Williamsburg, VA: Institute of Early American History and University of North Carolina Press, 1994), 278–327. Cooper used listings in Philadelphia City directories to make this estimate.

41. Anthony F. C. Wallace, *Rockdale: The Growth of an American Village in the Early Industrial Revolution* (New York: Norton, 1978), 214–16.

42. Cooper, "Patent Transformation," 305–6; quoted in McMahon and Morris, *Technology in Industrial America*, ix.

43. Sinclair, *Philadelphia's Philosopher Mechanics*, 135–69; Reynolds, *Stronger Than a Hundred Men*, 252–55; Bruce Sinclair, "Science, Technology, and the Franklin Institute," in *Pursuit of Knowledge in the Early American Republic: American Scientific and Learned Societies from Colonial Times to the Civil War*, ed. Alexandra Oleson and Sanborn C. Brown (Baltimore: Johns Hopkins University Press, 1974), 194–207. Quoted in Sinclair, *Philadelphia's Philosopher Mechanics*, 147.

44. Quoted in Sinclair, *Philadelphia's Philosopher Mechanics*, 141.

45. Franklin Institute, "Report of the Committee of the Franklin Institute of Pennsylvania, Appointed May 1829, to Ascertain by Experiment the Value of Water as a Moving Power," *Journal of the Franklin Institute* 11 (1831): 145–54; 12 (1831): 73–89, 147–53, 221–30, 296–305, 367–73; 13 (1831): 31–39, 153–57, 295–303, 370–77; 14 (1832): 10–16, 294–302, 366–71; 31 (1841): 145–54, 217–24, 289–96, 361–69; 32 (1841): 1–8.

46. Edward W. Constant II, *The Origins of the Turbojet Revolution* (Baltimore: Johns Hopkins University Press, 1980); and Edwin T. Layton Jr., "Scientific Technology, 1845–1900: The Hydraulic Turbine and the Origins of American Industrial Research," *Technology and Culture* 20 (1979): 64–89; Sinclair, *Philadelphia's Philosopher Mechanics*, 240.

47. Committee on Water Power, *Journal of the Franklin Institute* 11 (1831): 152.

48. "In order that this measure of effect may be accurate, the friction and inertia of the machine must be considered. The friction was carefully ascertained by experiment and the proper allowance made for it, as will appear in the course of this report. Any resistance from inertia was avoided by causing the wheel, and of course the weight raised by it, to

move, before beginning an experiment, with a velocity which would remain constant during its progress." Report of the Committee on Water Power, *Journal of the Franklin Institute* 11 (1831): 147.

49. Ibid., 87.

50. Personal conversation with Bruce Sinclair.

51. Velocity of the wheel was figured, to see if it influenced the effect (it was determined that it did not). Velocity was not used in other calculations.

Two • The Effects of Control

1. The term *efficiency* had no exact counterpart in French. The term *perfectionnement* encompassed not only the mathematical ratio of output to input but also broader issues of the social and economic performance of machines. The French term *efficacité* does not capture the concept as a comparative measurement, denoting instead planned effectiveness more generally. *Rendement*, or performance, was often the closest equivalent.

2. Ken Alder, *Engineering the Revolution: Arms and Enlightenment in France, 1763–1815* (Princeton: Princeton University Press, 1997), 315–17.

3. Claudine Fontanon, "Conviction républicaine pour une fondation," in *Le Conservatoire National des Arts et Métiers au coeur de Paris*, ed. Claudine Fontanon (Paris: Conservatoire National des Arts et Métiers, 1994), 60–68.

4. Frederick B. Artz, *The Development of Technical Education in France, 1500–1850* (Cambridge, MA: MIT Press and Society for the History of Technology, 1966), 143–50.

5. Convention Nationale, Instruction Publique, *Rapport sur l'etablissement d'un Conservatoire des Arts et Métiers par Gregoire. Seance du 8 vendemaire an 3*, Paris, 1793, 13–14, trans. James M. Edmonson, *From Mécanicien to Ingénieur: Technical Education and the Machine Building Industry in Nineteenth-Century France* (New York: Garland, 1987), 55.

6. Fontanon, "Conviction républicaine," 62.

7. V. Leblanc, *Nouveau système complet de filature de coton, usité en Angleterre et importé en France par la compagnie établie à Ourscamp, prés Compiègne* (Paris: Bachelier, 1828), xi; Margaret Bradley and Fernand Perrin, "Charles Dupin's Study Visits to the British Isles, 1816–1824," *Technology and Culture* 32 (1991): 47–68. Leblanc's work was incorrectly catalogued at the Bibliothèque Nationale under V. Leblanc; his widow had continued to operate his presses and workshop after his death, issuing engravings signed "V. Leblanc," for Veuve Leblanc; see Louis André, "César Nicolas Leblanc et le Dessin de Machines," *Les Cahiers d'Histoire du CNAM* 4 (1994): 71–92.

8. Robert Fox, "Science, the University, and the State in Nineteenth-Century France," in *Professions and the French State, 1700–1900*, ed. Gerald L. Geison (Philadelphia: University of Pennsylvania Press, 1984), 119.

9. Report to Interior Minister, 1814, personnel dossier, Christian, G.-J., ser. F12 Conservatoire des Arts et Métiers, Conseil de perfectionnement, personnel, matériel, inventions, an IV–1879, file F12 4862, Archives Nationales, Paris; biographical notices for G.-J. Christian, 1880, 1903, doc. no. 2769, Centre d'Histoire des Techniques du Conservatoire National des Arts et Métiers, Paris; Antoine Picon, "Christian, Gérard-Joseph (1778–1832), Directeur du Conservatoire (1816–1831)," *Les Professeurs du Conservatoire Nationale*

des Arts et Métiers: dictionnaire biographique 1794–1955 (Paris: Institut National de Recherche Pédagogique, 1994), 332–36; Anatole de Monzie, *Le Conservatoire du peuple* (Paris: Sedes, 1948), 57.

10. Charles Ballot, *L'Introduction du machinisme dans l'industrie française* (Lille: O. Marquant; Paris: F. Rieder and Co., 1923), 128.

11. Robert Fox, "Education for a New Age," in *Artisan to Graduate: Essays to Commemorate the Foundation in 1824 of the Manchester Mechanics' Institute*, ed. Donald S. L. Cardwell (Manchester: Manchester University Press, 1974), 24–38.

12. John Hubbel Weiss, *The Making of Technological Man: The Social Origins of French Engineering Education* (Cambridge, MA: MIT Press, 1982).

13. Edmonson, *From Mécanicien to Ingénieur*, 188–95; Fox, "Education for a New Age," 31–32; D. S. L. Cardwell, *From Watt to Clausius: The Rise of Thermodynamics in the Early Industrial Age* (Ithaca, NY: Cornell University Press, 1971), 191–201.

14. Fox, "Education for a New Age," 29–30; Andre Jardin and Andre-Jean Tudesq, *Restoration and Reaction, 1815–1848* (Cambridge: Cambridge University Press, 1983), 27–31.

15. Francis Démier, "Le Moment libéral de l'économie politique," in Fontanon, *Le Conservatoire National des Arts et Métiers*, 98.

16. Liliane Hilaire-Pérez, "Invention and the State in 18th Century France," *Technology and Culture* 32 (1991): 911–31; Roger Hahn, *The Anatomy of a Scientific Institution: The Paris Academy of Sciences, 1666–1803* (Berkeley: University of California Press, 1971), 66–68; Christine MacLeod, "The Paradoxes of Patenting: Invention and Its Diffusion in 18th and 19th Century Britain, France, and North America," *Technology and Culture* 32 (1991): 892–93.

17. Fontanon, "Conviction républicaine," 62.

18. Ferdinand-Dreyfus, *Un Philanthrope d'autrefois: La Rochefoucauld-Liancourt, 1747-1827* (Paris: Librairie Plon, 1903), 416–19; Christian to director of the Louvre, letter of August 27, 1819, manuscript no. 414, Library of the Conservatoire National des Arts et Métiers.

19. G.-J. Christian, *Vues sur le système général des opérations industrielles, ou plan de Technonomie* (Paris: Huzard and Courcier, 1819), 64–65.

20. Dreyfus, *Un Philanthrope d'autrefois*, 414–15; Picon, "G.-J. Christian."

21. Gérard-Joseph Christian, "Notice sur le Conservatoire Royal des Arts et Métiers," *Catalogue général des collections du Conservatoire Royal des Arts et Métiers* (Paris: Huzard, 1818), i–xx, i–ii, xviii–xix.

22. Dreyfus, *Un Philanthrope d'autrefois*, 301.

23. Gérard-Joseph Christian, *Instruction pour les gens de la campagne sur la manière de préparer le lin et le chanvre sans rouissage* (Paris: Huzard, 1818); *Descriptions des machines et procédés spécifiés dans les brevets d'invention, de perfectionnement et d'importation dont la durée est expirée* (Paris: Huzard, 1818–30); *Plan de Technonomie; Traité de mécanique industrielle ou exposé de la science de la mécanique déduite de l'expérience et de l'observation* (Paris: Bachelier, 1822–25).

24. *L'Industriel* 1, no. 1 (1826): 1.

25. Dreyfus, *Un Philanthrope d'autrefois*, 414; M. Leblanc, "Cours de dessin appliquée aux machines," *L'Industriel* 1, no. 2 (1926): 56–58.

26. Christian, *Plan de Technonomie*, 72; Christian, *Traité de mécanique industrielle*, 1:iv.

27. Edmonson, *From Mécanicien to Ingénieur*, 3, 310–25; Jennifer Karns Alexander, "The Line between Potential and Working Machines: César Nicolas Leblanc and Patent Engravings, 1811–1835," *History and Technology* 15 (1999): 175–212.

28. Roger Magraw, *A History of the French Working Class* (Oxford: Blackwell, 1992), 40–43.

29. William Hamilton Sewell, *Work and Revolution in France: The Language of Labor from the Old Regime to 1848* (Cambridge: Cambridge University Press, 1980).

30. Jacques Étienne Bédé, "The Life and Military Career of J. E. Bédé, Written by Himself," in *The French Worker: Autobiographies from the Early Industrial Era*, ed. and trans. Mark Traugott (Berkeley: University of California Press, 1993), 69, 71.

31. Gérard Noriel, *Les Ouvriers dans la société française XIXe–XXe siècle* (Paris: Éditions du Seuil, 1986), 43–55.

32. William H. Sewell Jr., "Artisans, Factory Workers, and the Formation of the French Working Class," in *Working-Class Formation: Nineteenth-Century Patterns in Western Europe and the United States,* ed. Ira Katznelson and Aristide R. Zolberg (Princeton: Princeton University Press, 1986), 45.

33. Jan Sebestik, "De la Technologie à la Technonomie: Gérard-Joseph Christian," *Cahiers S.T.S.* 2 (1984): 56–69; Jacques Guillerme and Jan Sebestik, "Les Commencements de la technologie," *Thales* 12 (1966): 58–62; Joost Mehrtens, "Technology as the Science of the Industrial Arts: Louis-Sébastien Lenormand (1757–1837) and the Popularization of Technology," *History and Technology* 18 (2002): 205.

34. Sebestik, "De la Technologie à la Technonomie."

35. Christian to Rochefoucauld, letter of July 2, 1817, manuscript no. 413, Library of the Conservatoire National des Arts et Métiers, Paris.

36. Christian, *Plan de Technonomie*, 69–70.

37. Ibid., 67–69.

38. Quoted in Sewell, "Artisans, Factory Workers," 46.

39. Ibid., 49.

40. Christian, *Plan de Technonomie*, 45–49.

41. Abbott Payson Usher, "The Textile Industry, 1750–1830," in *Technology in Western Civilization*, ed. Melvin Kranzberg and Carroll Pursell Jr. (New York: Oxford University Press, 1967), 1:232–38.

42. Christian, *Plan de Technonomie*, 25–26, 64–65, 15.

43. Ibid., 12.

44. Ibid., 14.

45. Ibid., 109–11, 26–27.

46. Ibid., 149–51, 104, 14, 99–106, quote on 150.

47. Ibid., 114–15; Christian, *Traité de mécanique industrielle*, 2:374.

48. Leblanc, "Machine a vapeur a haute pression, avec detente variable, de la force de 25 chevaux, construite par M. Pauwels, à Paris"; and "Turbine hydraulique et tender construite par M. Fourneyron," *Receuil des Machines, Instrumens et Appareils Qui Servent à l'Économie Rurale et Industrielle* (Paris, 1826), 3:3, 9–10.

49. Leblanc, "'Turbine,' par M. Fontaine, mécanicien à Chartres," *Receuil des Machines,* 4:7.

50. *Catalogue des spécifications de tous les principles, moyens et procédés pour lesquels il a été pris des brevets d'invention, de perfectionnement et d'importation, depuis le 1er juillet 1791, epoque de la mise a execution des lois des 7 janvier et 25 mai precedens, jusqu'au 1er juillet 1825* (Paris: Anthelme Boucher, 1826), 250–54. Forty-three patents were issued; sixteen were for *perfectionnement,* some also for importation or invention.

51. "Pour une machine à vapeur, a mouvement parallele, sans balancier, au sieur John Leberecht-Steinhauser, de Londres, 12 juillet 1822," *Descriptions des machines et procédés, consignés dans les brevets d'invention, de perfectionnement et d'importation* (Paris, 1833), 24:139.

52. "Pour un moyen de communiquer la vapeur dans une machine a haute pression, de manière a economiser le combustible, plus que ne l'ont fait MM. Aitken et Steel, au sieur Hall fils (Edward) de Darford, en Angleterre," *Descriptions des machines,* 23:153, pl. 20.

53. Jean-Christian Dietz, "Machine rotative à vapeur," n.d., indexed with steam engines before 1840, Central Archive, Institut National de la Propriété Industrielle, Compiègne.

54. See Coriolis, *De calcul de l'effet des machines* (Paris: Carilian-Goeury, 1829), iii–iv, 14–17; and G.-J. Christian, "De la manière d'appreciés en général la force dans les travaux industriels," *L'Industriel* 1, no. 2 (1826): 65–92.

55. Christian, *Traité de mécanique industrielle,* 2:396.

56. Ibid., 1:14.

57. Ibid., 1:13.

58. Theodore M. Porter, *Trust in Numbers: The Pursuit of Objectivity in Science and Public Life* (Princeton: Princeton University Press, 1995), 55.

59. Chaptal, *De l'industrie française* (1819; rpt., Paris: Imprimerie Nationales Éditions, 1993), 263–64.

60. Christian, *Traité de mécanique industrielle,* 1:2. The translation is Edmonson's, in *From Mécanicien to Ingénieur,* 193; the same passage also appears in Christian, *Plan de Technonomie,* 69–70.

61. Christian, *Traité de mécanique industrielle,* 3:18.

62. Ibid., 3:19.

63. Ibid., 2:395.

64. Daryl M. Hafter, "Measuring Cloth by the Elbow and a Thumb: Resistance to Numbers in France of the 1780s," in *Cultures of Control,* ed. Miriam R. Levin (Amsterdam: Harwood Academic Publishers, 2000), 73–78; Magraw, *History of the French Working Class,* 40–43.

65. Leblanc, *Nouveau système complet de filature de coton,* viii–xii.

Three • Economy of Nature

1. J. B. Bury, *The Idea of Progress: An Inquiry into Its Origin and Growth* (New York: Macmillan, 1932).

2. Both Darwin's and Marshall's thought evolved considerably over their lifetimes, Darwin's after the *Origin* and Marshall's leading up to the last edition of his most famous work. On changes Darwin made to successive editions of the *Origin,* see Janet Browne, *Charles Darwin: The Power of Place* (New York: Knopf, 2003). For changes in Marshall's thought,

see C. W. Guillebaud, "The Evolution of Marshall's *Principles of Economics*," *Economic Journal* 52 (December 1942): 330–49. Marshall amended book 4 of the *Principles* most heavily; the changes dealt principally with utility, consumer's surplus, and the theory of quasirents. Guillebaud produced a useful annotated variorum edition of the *Principles;* see Alfred Marshall, *Principles of Economics*, 9th ed. (variorum) (London: Macmillan, for the Royal Economic Society, 1961); vol. 1 is the text of the 8th ed., vol. 2 the annotations.

3. Ivor Grattan-Guinness, "Work for the Workers: Advances in Engineering Mechanics and Instruction in France, 1800–1830," *Annals of Science* 41 (1984): 1–33.

4. E. Mendoza, ed., *Reflections on the Motive Power of Fire, Sadi Carnot, and Other Papers on the 2nd Law of Thermodynamics* (New York: Dover, 1960), 12.

5. Philip Mirowski, *More Heat Than Light: Economics as Social Physics: Physics as Nature's Economics* (Cambridge: Cambridge University Press, 1989), 23–26.

6. Ben Marsden, "Engineering Science in Glasgow: Economy, Efficiency and Measurement as Prime Movers in the Differentiation of an Academic Discipline," *British Journal for the History of Science* 25 (1992): 342. Italics in original.

7. Quoted in ibid., 328.

8. Thomas S. Kuhn, "Energy Conservation as an Example of Simultaneous Discovery," in *Critical Problems in the History of Science,* ed. Marshall Clagett (Madison: University of Wisconsin Press, 1959), 321–56.

9. James Robert Napier, Walter Neilson, and W. J. M. Rankine, "Report on the Progress and State of Applied Mechanics," *Proceedings of the Glasgow Philosophical Society* 4 (1855–60): 225.

10. Norton Wise and Crosbie Smith, "Measurement, Work, and Industry in Lord Kelvin's Britain," *Historical Studies in the Physical and Biological Sciences* 17 (1986): 147–73; and Wise and Smith, "Work and Waste: Political Economy and Natural Philosophy in Nineteenth Century Britain," *History of Science*, pt. 1, 27 (1989): 263–301; pt. 2, 27 (1989): 391–449; pt. 3, 28 (1990): 222–61; see pt. 3.

11. See Marsden, "Engineering Science in Glasgow."

12. Napier, Neilson, and Rankine, "Report," 208.

13. Rankine, quoted in Marsden, "Engineering Science in Glasgow," 328, 330; Marsden, "Engineering Science in Glasgow," 342.

14. Philip H. Rehbock, "John Fleming (1785–1857) and the Economy of Nature," in *From Linnaeus to Darwin: Commentaries on the History of Biology and Geology,* ed. Alwyne Wheeler and James H. Price (London: Society for the History of Natural History, 1985), 133.

15. Frank N. Edgerton, "Changing Concepts of the Balance of Nature," *Quarterly Review of Biology* 48 (1973): 322–50; see also Michael Ruse, *The Evolution-Creation Struggle* (Cambridge: Harvard University Press, 2005).

16. Wise and Smith, "Work and Waste," pt. 1, 267.

17. Maxine Berg, *The Machinery Question and the Making of Political Economy, 1815–1848* (Cambridge: Cambridge University Press, 1981), 108.

18. C. W. Guillebaud, "The Evolution of Marshall's *Principles of Economics*."

19. Alfred Marshall, *Principles of Economics: An Introductory Volume,* 8th ed. (1920; rpt., London: Macmillan, 1946), 2–4.

20. Charles Darwin, *On the Origin of Species by Means of Natural Selection, or the Preservation of Favoured Races in the Struggle for Life,* facsimile of 1st ed. (1859; rpt., Cambridge: Harvard University Press, 1964), 73, 490. On changes Darwin made to successive editions of the *Origin,* see Browne, *Charles Darwin: The Power of Place.*

21. Ibid., 467.

22. Ibid., 467, 43, 61, 109.

23. Ibid., 242.

24. Ibid., 237.

25. Ibid., 237, 238, 239, 241.

26. Ibid., 241–42.

27. Wise and Smith, "Work and Waste," pt. 2; on Darwin and Smith, see Silvan S. Schweber, "Darwin and the Political Economists: Divergence of Character," *Journal of the History of Biology* 13 (1980): 195–289.

28. Janet Browne, "Darwin's Botanical Arithmetic and the 'Principle of Divergence,' 1854–1858," *Journal of the History of Biology* 13 (1980): 88.

29. Darwin, *Origin,* 83–84.

30. Ibid., 467–68.

31. Ibid., 252, 451, 104.

32. Schweber, "Darwin and the Political Economists."

33. Darwin, *Origin,* 113.

34. Ibid., 114.

35. Browne, "Darwin's Botanical Arithmetic."

36. Darwin, *Origin,* 469, 489, 82, 83.

37. See Wise and Smith, "Work and Waste," pt. 2.

38. Darwin, *Origin,* 84, 67.

39. Robert M. Young, *Darwin's Metaphor: Nature's Place in Victorian Culture* (Cambridge: Cambridge University Press, 1985), 121.

40. See Browne, *Charles Darwin: The Power of Place,* on Darwin's adoption of Herbert Spencer's term *survival of the fittest* in response to criticism of his alleged personification of natural selection.

41. Rehbock, "John Fleming," 133.

42. Robert Chambers, *Vestiges of the Natural History of Creation,* 7th ed. (London: John Churchill, 1860), 279.

43. Donald Worster, *Nature's Economy: A History of Ecological Ideas* (Cambridge: Cambridge University Press, 1994), 187.

44. Darwin, *Origin,* 122, 201.

45. Dov Ospovat, *The Development of Darwin's Theory: Natural History, Natural Theology, and Natural Selection, 1838–1859* (Cambridge: Cambridge University Press, 1981), 226–28.

46. Joseph A. Schumpeter, "Alfred Marshall's *Principles:* A Semi-Centennial Appraisal," *American Economic Review* 31 (1941): 237.

47. Margaret Schabas, "The Greyhound and the Mastiff: Darwinian Themes in Mill and Marshall," in *Natural Images in Economic Thought: Markets Read in Tooth and Claw,* ed. Philip Mirowski (Cambridge: Cambridge University Press, 1994), 322–35.

48. Guillebaud, "The Evolution of Marshall's *Principles of Economics*."

49. Jurg Niehans, *A History of Economic Theory: Classic Contributions, 1720–1980* (Baltimore: Johns Hopkins University Press, 1990), 236–47.

50. George J. Stigler, "Marshall's *Principles* after Guillebaud," in *Alfred Marshall: Critical Assessments,* ed. John Cunningham Wood (London: Croom Helm, 1982), 2:223; in the same collection see also the memoir of his student, John Maynard Keynes, "Alfred Marshall, 1842–1924," 7–65.

51. Ernesto Screpanti and Stefano Zamagni, *An Outline of the History of Economic Thought,* trans. David Field (Oxford: Clarendon Press, 1993), 179.

52. Marshall, *Principles,* xiv–xv.

53. Margaret Schabas, *The Natural Origins of Economics* (Chicago: University of Chicago Press, 2005), 150–58.

54. Marshall, *Principles,* 240.

55. Ibid., 318.

56. Ibid., 265, 278, 240.

57. Ibid., 291–313.

58. Ibid., 253, 205–6.

59. Ibid., 206.

60. Peter Groenewegen, A *Soaring Eagle: Alfred Marshall, 1841–1924* (Aldershot: Edward Elgar, 1995), 171.

61. Ibid., 200–201.

62. "Notes on Efficiency of Labor," file 4/25, Marshall Papers, Marshall Library of Economics, Cambridge University, England.

63. "Notes on Efficiency of Labor," Marshall Papers, Marshall Library of Economics, f. 7.

64. "Notes on Efficiency of Labor," f. 8–20, 13, 15–16.

65. Groenewegen, *Soaring Eagle,* 193–203.

66. Alfred Marshall, *The Correspondence of Alfred Marshall, Economist,* vol. 1: *Climbing, 1868–1890,* ed. John K. Whitaker (Cambridge: Cambridge University Press, 1996), 1:50.

67. Ibid., 1:49.

68. Ibid., 1:50–51.

69. See Camille Limoges and Claude Ménard on the possibility that Marshall introduced his theory of the representative firm to step back from the biological analogies that suggested that, on the principle of competitive exclusion, an economy dominated by monopolies would inevitably result. An analysis of the role of efficiency in Marshall suggests a similar conclusion: efficiency, which both grew out of and further stimulated the differentiation of tasks within the control of business management, might also bring monopoly domination as its inevitable result. Limoges and Ménard, "Organization and the Division of Labor: Biological Metaphors at Work in Alfred Marshall's *Principles of Economics,*" in Mirowski, *Natural Images in Economic Thought,* 349.

70. Marshall, *Principles,* 206–7, 213.

71. Ibid., 293, 312.

72. Ibid., 193–203.

73. Ibid., 241–43.

74. Groenewegen, *Soaring Eagle,* 200–201.

75. Jacob Viner, "Marshall's Economics in Relation to the Man and His Times," *American Economic Review* 31 (1941): 223–35.

Four • Balance and Transformation

1. H. L. Gantt, "Compensation of Workmen and Efficiency of Operation: IV. Fixing Habits of Industry," *Engineering Magazine* 39 (1910): 172; L. P. Alford, *Henry Laurence Gantt: Leader in Industry* (New York: American Society of Mechanical Engineers, 1934), 35–45, 107–37; Samuel Haber, *Efficiency and Uplift: Scientific Management in the Progressive Era* (Chicago: University of Chicago Press, 1964), 44.

2. "Efficiency Question Box," *Independent* 82 (1915): 86.

3. Richard Hofstadter, *The Age of Reform, from Bryan to F.D.R.* (New York: Vintage, 1955), 10.

4. Daniel T. Rodgers, "In Search of Progressivism," *Reviews in American History* 10 (December 1982): 123, 126.

5. Historians have generally located the United States's and Britain's particular zeal for efficiency in the decades leading up to World War I; reform efforts centering on efficiency in France and Germany had significant presence after the war and through the 1920s; see G. R. Searle, *The Quest for National Efficiency: A Study in British Politics and Political Thought, 1899–1914* (1971; rpt., London: Ashfield Press, 1990); Sheila Faith Weiss, *Race Hygiene and National Efficiency: The Eugenics of Wilhelm Stallmayer* (Berkeley: University of California Press, 1987); and Mary Nolan, *Visions of Modernity: American Business and the Modernization of Germany* (New York: Oxford University Press, 1994).

6. Haber, *Efficiency and Uplift,* 55–56.

7. James Quigel Jr., "The Business of Selling Efficiency: Harrington Emerson and the Emerson Efficiency Engineers, 1900–1930" (Ph.D. diss., University of Pennsylvania, 1992), 368–69.

8. J. G. Horner, "Dictionary of Traditional Terms," *Dictionary of Terms Used in the Theory and Practice of Mechanical Engineering,* 8th ed. (1936; rpt., London: Technical Press, 1960), 125, 228.

9. Donald Cardwell, *The Norton History of Technology* (New York: Norton, 1995), 85, 88–89.

10. R. Caird, "The Comparative Efficiency of Marine Engines and Turbines: A Comparison Based on the Performances of the 'Kaiser Wilhelm II' and the 'Lusitania,'" *Engineering Magazine* 36 (1909): 696–98.

11. Frank C. Wagner, "The Possibilities of the Gasoline Turbine: An Estimate of the Possible Combined Efficiency of Turbine and Compressor," *Engineering Magazine* 36 (1909): 708–10; Walter N. Polakov, "Efficiency in the Purchase of Fuel for Power Generation," *Engineering Magazine* 38 (1909): 217, 218, 219.

12. "The Basis of Efficiency," *Engineering Record* 65 (1912): 1; Bruce Sinclair, *The American Society of Mechanical Engineers: A Centennial History of the American Society of Mechan-*

ical Engineers, 1880–1980 (Toronto: University of Toronto Press, 1980); Haber, *Efficiency and Uplift*, ix.

13. Monte A. Calvert, *The Mechanical Engineer in America, 1830–1910: Professional Cultures in Conflict* (Baltimore: Johns Hopkins University Press, 1967), 43–62; Amy Slaton, *Reinforced Concrete and the Modernization of American Building, 1900–1930* (Baltimore: Johns Hopkins University Press, 2001), 240.

14. Robert L. Streeter, "The Internal-Combustion Engine in Modern Practice," *Engineering Magazine* 42 (1911): 204.

15. Robert L. Streeter, "The Internal-Combustion Engine in Modern Practice. V. Present Practice in Germany," *Engineering Magazine* 43–44 (1912): 735.

16. Heldt, *High-Speed Combustion Engines: Design: Production: Tests*, 11th ed. of *The Gasoline Motor* (Nyack, NY: P. M. Heldt, 1941), 665–72. Brake horsepower was converted to heat using the equivalent 1 hp = 2,545 btu/hr.

17. Thomas L. Hankins and Robert J. Silverman, *Instruments and the Imagination* (Princeton: Princeton University Press, 1995), 128–29.

18. P. M. Heldt, *The Gasoline Motor* (New York: Horseless Age, 1911), 435, 448.

19. Heldt, *High-Speed Combustion Engines*, 628. The high heat value would have been calculated using a Junker's calorimeter.

20. "The Testing of Gasoline Engines," *Horseless Age* 15 (1905): 459–60.

21. Haber, *Efficiency and Uplift*, 33; Polakov, "Efficiency in the Purchase of Fuel for Power Generation."

22. Harrington Emerson, "Efficiency as a Basis for Operation and Wages. VIII. The Efficiency System in Operation," *Engineering Magazine* 36 (1909): 815; Emerson, "Efficiency as a Basis for Operation and Wages. IV. Standards; Their Relations to Organization and to Results," *Engineering Magazine* 36 (1908): 37, 38, 33; "Harrington Emerson," *National Cyclopedia of American Biography* (New York: James T. White, 1916), 81–82; Harrington Emerson, "Discussion," *Transactions of the American Society of Mechanical Engineers* 25 (1904): 73.

23. Harrington Emerson, "Efficiency as a Basis for Operations and Wages. VI. The Modern Theory of Cost Accounting," *Engineering Magazine* 36 (1908): 337.

24. Emerson, "The Efficiency System in Operation," 815, 822.

25. Calvert, *Mechanical Engineer in America*; and Daniel Hovey Calhoun, *The American Civil Engineer: Origins and Conflict* (Cambridge, MA: Technology Press, MIT, 1960), 243.

26. Quigel, "Business of Selling Efficiency," 7, 125–215.

27. Charles Buxton Going, "Methods of the Santa Fe. Efficiency in the Manufacture of Transportation. I. Peculiar Problems of the Road, and Their Solutions," *Engineering Magazine* 36 (1909): 910.

28. Charles Buxton Going, "The Methods of the Santa Fe: Efficiency in the Manufacture of Transportation. III. Manufacturing Policies for the Economical Maintenance of Motive Power," *Engineering Magazine* 37 (1909): 238; Daniel T. Rodgers, *The Work Ethic in Industrial America, 1850–1920* (Chicago: University of Chicago Press, 1978), 56–57; Emerson, "The Efficiency System in Operation," 822.

29. Charles Buxton Going, "Methods of the Santa Fe. Efficiency in the Manufacture of

Transportation. IV. Administration, Supervision, and Extension of the Bonus System," *Engineering Magazine* 37 (1909): 339.

30. Charles Buxton Going, "Methods of the Santa Fe. Efficiency in the Manufacture of Transportation. V. The Apprentice System and Relations with Employees," *Engineering Magazine* 37 (1909): 541–64.

31. Alford, *Henry Laurence Gantt;* H. L. Gantt, "Compensation of Workmen and Efficiency of Operation. III. Task and Bonus," *Engineering Magazine* 39 (1910): 17–23; "IV. Fixing Habits of Industry," *Engineering Magazine* 39 (1910): 161–75; "V. Prices and Profits. A New Light on the Cost of Living," *Engineering Magazine* 39 (1910): 331–39; *Work, Wages and Profits: Their Influence on the Cost of Living* (New York: Engineering Magazine, 1910); and *Industrial Leadership: Page Lectures, 1915, before the Sheffield Scientific School, Yale University* (New Haven: Yale University Press, 1916).

32. Gantt, "Fixing Habits of Industry," 167.

33. Frank Luther Mott, *A History of American Magazines, 1850–1865* (Cambridge: Harvard University Press, 1938), 379.

34. Ibid., 367, 378–79; Frank Luther Mott, *A History of American Magazines, 1865–1885* (Cambridge, MA: Belknap, 1957), 435; Haber, *Efficiency and Uplift,* 60.

35. "Social Efficiency," *Independent* 71 (1911): 1103–4; "Efficiency Program," *Independent* 71 (1911): 739–40; "The Cost of Efficiency," *Independent* 77 (1914): 255; John Whiteclay Chambers II, *The Tyranny of Change: America in the Progressive Era, 1890–1920,* 2nd ed. (New Brunswick: Rutgers University Press, 2000), 6.

36. Edward Earle Purinton, *Efficient Living* (New York: Robert H. McBride & Co., 1915).

37. See the collection edited by Daniel Nelson, *A Mental Revolution: Scientific Management since Taylor* (Columbus: Ohio State University Press, 1992).

38. Edward Earle Purinton, "What Is Efficiency?" *Independent* 80 (1914): 320–23.

39. "The Ambassador," *Independent* 81 (1915): 185.

40. "Efficiency Question Box," *Independent* 81 (1915): 34.

41. "Efficiency Question Box," *Independent* 80 (1914): 480.

42. "Efficiency Question Box," *Independent* 81 (1915): 103; Edward Earle Purinton, "Work and Efficiency," *Independent* 80 (1914): 503.

43. Edward Earle Purinton, "Efficiency and Life," *Independent* 80 (1914): 323.

44. Edward Earle Purinton, "Efficiency Money Quiz," *Independent* 83 (1915): 118.

45. Edward Earle Purinton, "Efficiency Food Quiz," *Independent* 82 (1915): 154.

46. Edward Earle Purinton, "Health Efficiency Gauge," *Independent* 82 (1915): 324.

47. Robert H. Wiebe, *The Search for Order, 1877–1920* (New York: Hill and Wang, 1967), 145.

48. "Efficiency Question Box," *Independent* 80 (1914): 480; 81 (1915): 103, 134, 260; 82 (1915): 206.

49. "Efficiency Question Box," *Independent* 81 (1915): 104, 340, 404.

50. Edward Earle Purinton, "Efficiency Work Scheme," *Independent* 80 (1914): 501.

51. "Efficiency Question Box," *Independent* 83 (1915): 244.

52. Purinton, "Work and Efficiency," 500.

53. "Efficiency Question Box," *Independent* 81 (1915): 260.

54. "Efficiency Question Box," *Independent* 82 (1915): 86; 83 (1915): 337.

Five • An Island of Mechanical Predictability

1. There was no exact counterpart in German for the English word *efficiency*. *Efficienz* is a more recent, postwar term. *Wirkungsgrad* expressed the technical measurement, especially in terms of energy use, while *Nutzeffekt* denoted a more general sense of usefulness. *Wirtschaftlichkeit* and *Liestungsfähigkeit* were also translated as *efficiency*, but their use differed in an important respect. *Wirtschaftlichkeit* was the more general term, describing economic efficiency; this was the sense in which a particular industry was efficient, for example, and this was the term used to describe the national benefits and rich rewards of efficiency. *Leistungsfähigkeit* was applied in a more technical sense, to specific machines or processes such as new Reichsbahn engines or the human metabolism. It connoted the capacity to perform and was the term used in discussions of human labor performance analyzed in this chapter.

2. Lisa Cartwright, *Screening the Body: Tracing Medicine's Visual Culture* (Minneapolis: University of Minnesota Press, 1993), 47–80; Erin O'Connor, *Raw Material: Producing Pathology in Victorian Culture* (Durham: Duke University Press, 2000), 7; Friedrich Engels, *The Condition of the Working Class in England,* ed. David McLellan (New York: Oxford University Press, 1993), 210.

3. Marcel Mauss, "Body Techniques," *Sociology and Psychology* (London: Routledge and Kegan Paul, 1979), 104; William H. McNeill, *Keeping Together in Time: Dance and Drill in Human History* (Cambridge: Harvard University Press, 1995), viii, 9–10, 101–50; Atina Grossman, "The New Woman and the Rationalization of Sexuality in Weimar Germany," in *Powers of Desire: The Politics of Sexuality,* ed. Ann Snitow, Christine Stansell, and Sharon Thompson (New York: Monthly Review Press, 1988), 164; see also Michel Foucault, *Discipline and Punish: The Birth of the Prison* (New York: Random House, 1977), 152.

4. Sheila Faith Weiss, *Race Hygiene and National Efficiency: The Eugenics of Wilhelm Schallmayer* (Berkeley: University of California Press, 1987); Paul Weindling, *Health, Race and German Politics between National Unification and Nazism, 1870–1945* (Cambridge: Cambridge University Press, 1989), 402; Erwin Baur, Eugen Fischer, and Fritz Lenz, *Human Heredity,* trans. Eden Paul and Cedar Paul (New York: Macmillan, 1931), 213–17. On the uniqueness of German eugenics, see Paul Weindling, "The Sonderweg of German Eugenics: Nationalism and Scientific Internationalism," *British Journal for the History of Science* 22 (1989): 321–33; Peter Weingart, "German Eugenics between Science and Politics," *Osiris* 5 (1989): 260–82; and Jonathan Harwood, "National Styles in Science: Genetics in Germany and the United States between the World Wars," *Isis* 78 (1987): 390–414.

5. Ludwig Ascher, "Der Mensch und die Rationalisierung: Bericht über die RKW-Tagung, 'Der Mensch und die Rationalisierung,'" *RKW-Nachrichten* 4 (1930): 48–50.

6. Georges Canguilhem, *On the Normal and the Pathological* (Dordrecht: D. Reidel, 1978), 240, 239, 146; see also Hans Wupper-Tewes, *Rationalisierung als Normalisierung: Betriebswissenschaft und betriebliche Leistungspolitik in der Weimarer Republik* (Münster: Westfälisches Dampfboot, 1995), 78–89.

7. Ludwig Ascher, "Der Einfluss technischer Verbesserungen auf die Gesundheit des Menschen, insbesondere des Arbeiters," in *Resoconti del Congresso,* vol. 2: *IIIo Congresso del*

Internazionale di Organizzazione Scientifica del Lavoro (Rome, 1927), 2:563–70; Anson Rabinbach, *The Human Motor: Energy, Fatigue, and the Origins of Modernity* (Berkeley: University of California Press, 1990), 84–119; Helmuth von Bracken, "Rationalisierung und Gesundheit: Gewerbehygienische Probleme auf dem III. Kongress für wissenschaftliche Organisation der Arbeit," *Zentralblatt für Gewerbehygiene und Unfallverhütung* 15 (1928): 1–3. In general German engineers and businessmen rejected as technocratic the scientific management of Taylor and colleagues, such as Frank Gilbreth and Lillian Gilbreth, preferring instead what many called "Fordism," which seemed to promise high productivity without intensified or exploitative work. Köttgen himself had been especially impressed with production methods in his visit to the Ford Motor Works and described them in 1925 in the widely read book *Das Wirtschaftliche Amerika* (Berlin, 1925). See Nolan, *Visions of Modernity: American Business and the Modernization of Germany* (New York: Oxford University Press, 1994), 30–57, 62; and Brian Price, "Frank and Lillian Gilbreth and the Motion Study Controversy, 1907–1930," in *A Mental Revolution: Scientific Management since Taylor*, ed. Daniel Nelson (Columbus: Ohio State University Press, 1992), 58–76, 63. Herbert Mehrtens and Richard Lindstrom have described the techniques used by the Gilbreths in their celebrated photographic studies; see Mehrtens, "Arbeit und Zeit, Körper und Uhr: Die Konstruktion von 'effektiver' Arbeit im Scientific Management," *Berichte zur Wissenschaftsgeschichte* 25 (2002): 121–36; and Lindstrom, "'They All Believe They Are Undiscovered Mary Pickfords': Workers, Photography, and Scientific Management," *Technology and Culture* 41 (2000): 725–51. For a study of the camera as a tool to bring the human body within the reach of industrial standards and norms, see Elspeth H. Brown, *The Corporate Eye: Photography and the Rationalization of American Commercial Culture, 1884–1929* (Baltimore: Johns Hopkins University Press, 2005).

8. Ascher, "Der Einfluss technischer Verbesserungen," 566–67.

9. Ibid., 563–64, 570; Ludwig Ascher, "Die Einwirkung der industriellen Entwicklung auf den Körper des Arbeiters," *Zentralblatt für Gewerbehygiene und Unfallverhütung* 16 (1929): 131–38; Ludwig Ascher, "Die Stellung des Sozialhygienikers zur modernen Industrieentwicklung," *Zentralblatt für Gewerbehygiene und Unfallverhütung* 16 (1929): 194–97. This meant that the workers had mastered their machines, Ascher believed; he did not address the different style of work an automatic loom required and what many have called the de-skilling that accompanied its use and became emblematic of weavers' transition from skilled artisans to machine tenders. D. A. Farnie describes how a weaver at an automatic loom became a "hopper-filler and piecer"; see Farnie, "The Textile Industry: Woven Fabrics" in the classic series edited by Charles Singer et al., *History of Technology* (London: Oxford University Press, 1958), 5:586. See also Gerhardt Hauptmann's classic play *Die Weber* (1893).

10. Franz Reuleaux, *Kinematics of Machinery: Outlines of a Theory of Machines*, trans. Alex Kennedy (London: Macmillan, 1876), 35, 87; Kees Gispen, *New Profession, Old Order: Engineers and German Society, 1815–1914* (Cambridge: Cambridge University Press, 1989), 115–21; Hans-Joachim Braun and Wolfhard Weber, "Ingenieurwissenschaft und Gesellschaftspolitik: Das Wirken von Franz Reuleaux," in *Wissenschaft und Gesellschaft: Beiträge zur Geschichte der Technischen Universität Berlin, 1879–1979*, ed. Reinhard Rürup (Berlin:

Springer Verlag, 1979), 1:285–300; and Alex Kennedy, preface to Reuleaux, *Kinematics of Machinery*, v–x.

11. Rabinbach, *Human Motor*, 38–44, 51–52, 271–88.

12. Ascher, "Der Einfluss technischer Verbesserungen"; Ludwig Ascher, "Bewegungsstudien zur Berufsauslese," *Beiheft Zentralblatt für Gewerbehygiene und Unfallverhütung* 9 (1928): 8–20; Ludwig Ascher, "Bewegungsstudien und Ermüdung," *Beiheft Zentralblatt für Gewerbehygiene und Unfallverhütung* 7 (1927): 66–73.

13. Martin Höfler-Waag, *Die Arbeits-und Leistungsmedizin im Nationalsozialismus von 1939–1945* (Husum: Matthiesen Verlag, 1994), 223. Hebestreit's commitment to using medicine for productivist ends became apparent under National Socialism. He was a leader in labor physiology during the Nazi years, having joined the party shortly after it seized power, and was an important member of the Nazi labor organization, the German Labor Front. Robert N. Proctor has described Hebestreit's view of occupational medicine: that it should try to narrow the gap between time of retirement and death to zero, the value of life (to the community) being equated with the ability to work; see Proctor, *The Nazi War on Cancer* (Princeton: Princeton University Press, 1999), 118–19.

14. Hermann Hebestreit, "Physiologische Grundlagen zur Frage der Arbeitsstellung und des Arbeitssitzes," *Zentralblatt für Gewerbehygiene und Unfallverhütung* 17 (1930): 5–7.

15. Ibid., 8–10; Hermann Hebestreit, "Die Arbeits-und Gewerbehygiene auf der Internationalen Hygieneausstellung Dresden 1930," *Reichsarbeitsblatt* 10 (1930), pt. 3: 129–35; Otto Bauer, *Kapitalismus und Sozialismus nach dem Weltkrieg: I. Rationalisierung, Fehrationalisierung* (Berlin: Büchergilde Gutenberg, 1931), 87–88.

16. Ascher, "Der Mensch und die Rationalisierung," 50; Ludwig Ascher, "Die Gesundheit des Arbeiters in der Weltkrise," *Zentralblatt für Gewerbehygiene und Unfallverhütung* 19 (1932): 205–7; Ascher attributed the concept to Hugo Münsterberg. See also Rabinbach, *Human Motor*, 191–92; and Joan Campbell, *Joy in Work, German Work: The National Debate, 1800–1945* (Princeton: Princeton University Press, 1989), 73–106.

17. Barbara Orland, "Effizienz im Heim: Die Rationalisierungsdebatte zur Reform der Hausarbeit in der Weimarer Republik," *Kultur und Technik* 4 (1983): 226.

18. Nolan, *Visions of Modernity*, 6.

19. Detlev J. Peukert, *The Weimar Republic* (New York: Hill and Wang, 1993), 118–19, 12–14.

20. Walter Meakin, *The New Industrial Revolution* (London: V. Gollancz, 1927), 17. Noted in Robert A. Brady, "The Meaning of Rationalization: An Analysis of the Literature," *Quarterly Journal of Economics* (May 1932): 526. The term had been used before the war, although not widely. See Wupper-Tewes, *Rationalisierung als Normalisierung*, 36–40. Quotation from J. Ronald Shearer, "Talking about Efficiency: Politics and the Industrial Rationalization Movement in Weimar Germany," *Central European History* 28 (1995): 489.

21. Charles Maier, "Zwischen Taylorismus und Technokratie," in *Die Weimarer Republik, Belagerte Civitas*, ed. Michael Stürmer (Königstein: Athenäum, 1980), 188–213; Wilfried Feldenkirchen, *Siemens, 1918–1945* (Columbus: Ohio State University Press, 1999), 663 n. 8; Carola Sachse, *Siemens, der Nationalsozialismus und die moderne Familie; Eine Unter-*

suchung zur socialen Rationalisierung in Deutschland im 20. Jahrhundert (Hamburg: Rausch und Röhring, 1990), 28, 95–242; Heidrun Homburg, *Rationalisierung und Industriearbeit: Arbeitsmacht-Management-Arbeiterschaft im Siemens-Konzern Berlin, 1900–1939* (Berlin: Haude & Spener, 1991), 586–661; Hinnenthal quoted in Nolan, *Visions of Modernity,* 71–72; Robert A. Brady, *The Rationalization Movement in German Industry: A Study in the Evolution of Economic Planning* (Berkeley: University of California Press, 1933), xx.

22. J. Ronald Shearer, "The Politics of Industrial Efficiency in the Weimar Republic: Technological Innovation, Economic Efficiency, and Their Social Consequences in the Ruhr Coal Mining Industry, 1918–1929" (Ph.D. diss., University of Pennsylvania, 1989), 1, 4; Homburg, *Rationalisierung und Industriearbeit,* 488–90; John Gillingham, *Industry and Politics in the Third Reich: Ruhr Coal, Hitler, and Europe* (New York: Columbia University Press, 1985), 7, 9; see also Thomas von Freyberg, *Industrielle Rationalisierung in der Weimarer Republik* (Frankfurt: Campus Verlag, 1989); Gunnar Stollberg, *Die Rationalisierungsdebatte 1908–1933* (Frankfurt: Campus Verlag, 1981); and Jürgen Kocka, *Industrial Culture and Bourgeois Society: Business, Labor, and Bureaucracy in Modern Germany* (New York: Berghahn Books, 1999).

23. Sachse, *Siemens,* 28–29; Homburg, *Rationalisierung und Industriearbeit,* 343.

24. Rudolf Hilferding, "*Die Aufgaben der Sozialdemokratie in der Republik,*" *Rede auf dem Parteitag der SPD zu Kiel,* trans. Patrick Goode (Berlin: Volksblatt, 1927); reprinted in *The Weimar Republic Sourcebook,* ed. Anton Kaes, Martin Jay, and Edward Dimendberg (Berkeley: University of California Press, 1994), 69–70.

25. Gunther Mai, "Politische Krise und Rationalisierungsdiskurs in den zwanziger Jahren," *Technikgeschichte* 62 (1995): 317–32; Nolan, *Visions of Modernity,* 71, 76–79, 132–35; Brady, "Meaning of Rationalization"; Brady, *Rationalization Movement;* Shearer, *Talking about Efficiency,* 485; Akos Paulinyi, "Massenproduktion und Rationalisierung," *Technikgeschichte* 56 (1989): 173–81; Campbell, *Joy in Work,* 131–57. See also Peter Berg, *Deutschland und Amerika 1918–1929: Über das deutsche Amerikabild der Zwaniger Jahre* (Lubeck: Matthiesen Verlag, 1963); Charles S. Maier, "Between Taylorism and Technocracy: European Ideologies and the Vision of Industrial Productivity in the 1920s," *Journal of Contemporary History* 5 (1970): 27–61; and Mary Nolan, "The Infatuation with Fordism: Social Democracy and Economic Rationalization in Weimar Germany," in *Chance und Illusion: Labour in Retreat,* ed. Wolfgang Maderthaner and Helmut Gruver (Vienna: Europaverlag, 1988), 151–84. Beyond mining operations in the Ruhr, where rationalization was most intense, the effects of the movement are difficult to assess. Productivity and wages rose, but so did unemployment, and general welfare seemed little improved.

26. Housework reform was also a significant component of the rationalization movement and the aspect that most closely affected personal and family life; see Orland, "Efficienz im Heim"; and Mary Nolan, "Housework Made Easy," in *Visions of Modernity,* 206–26.

27. Shearer, "Talking about Efficiency," 501; Detlev J. K. Peukert, *The Weimar Republic: The Crisis of Classical Modernity* (New York: Hill and Wang, 1989), 129–46; Young-Sun Hong, *Welfare, Modernity, and the Weimar State, 1919–1933* (Princeton: Princeton University Press, 1998), 44–74, 181–202; Carola Sachse, *Betriebliche Sozialpolitik als Familienpolitik in der Weimarer Republik und im Nationalsozialismus: Mit einer Fallstudie über die Firma*

Siemens, Berlin (Hamburg: Hamburger Institut für Sozialforschung, 1987); Gunther Mai, "Politische Krise," 330–32.

28. Brady, *Rationalization Movement,* 49–50.

29. Ibid., 172–80; Homburg, *Rationalisierung und Industriearbeit,* 265–68.

30. Shearer, *Politics of Industrial Efficiency,* 392; Eve Rosenhaft, *Beating the Fascists? The German Communists and Political Violence, 1929–1933* (Cambridge: Cambridge University Press, 1983), 185; Eva Cornelia Schöck, *Arbeitslosigkiet und Rationalisierung* (Frankfurt: Campus Verlag, 1977), 153–80; Nolan, *Visions of Modernity,* 169–72.

31. Hans-Ulrich Thamer, *Verführung und Gewalt: Deutschland, 1933–1945* (Berlin, 1986), 470, table in Feldenkirchen, *Siemens,* 78; Klaus P. Fischer, *Nazi Germany: A New History* (New York: Continuum, 1995), 216; Peukert, *Weimar Republic,* 252; Peter D. Stachura, "The Development of Unemployment in Modern German History," in *Unemployment and the Great Depression in Weimar Germany,* ed. Peter D. Stachura (New York: St. Martin's Press, 1986), 14.

32. Feldenkirchen, *Siemens,* 96–97, 128, 228–30; Dietrich Orlow, *A History of Modern Germany, 1871 to Present,* 3rd ed. (Englewood Cliffs, NJ: Prentice Hall, 1995), 167–69; Peukert, *Weimar Republic,* 253–55; Michael Burleigh, *The Third Reich: A New History* (New York: Hill and Wang, 2000), 125–27; David Crew, *Germans on Welfare: From Weimar to Hitler* (New York: Oxford University Press, 1998), 157–65.

33. Stachura, "Development of Unemployment," 13; Hans Mommsen, *The Rise and Fall of Weimar Democracy* (Chapel Hill: University of North Carolina Press, 1996), 267–317; Peukert, *Weimar Republic,* 258–69; David Abraham, *The Collapse of the Weimar Republic: Political Economy and Crisis,* 2nd ed. (New York: Holmes and Meier, 1986), 37, 281–303; Fischer, *Nazi Germany,* 218–24; and Koppel Pinson, *Modern Germany,* 2nd ed. (New York: Macmillan, 1966), 452–53, 476–78.

34. Nolan, *Visions of Modernity,* 230–32; Peukert, *Weimar Republic,* 112–17.

35. "Mittelversorgung des RKW," *RKW-Jahresbericht 1930* (Berlin: Reichskuratorium für Wirtschaftlichkeit [RKW], 1930), 27–28; "Mittelversorgung des RKW," *RKW-Jahresbericht 1931* (Berlin: RKW, 1931), 31; Nolan, *Visions of Modernity,* 232, 137.

36. "Arbeiten des RKW im Berichtsjahr 1930–31," *RKW-Jahresbericht 1930* (Berlin: RKW, 1931), 28–30.

37. "RKW and Rationalisierung," *RKW-Nachrichten* 5 (1931): 112; "Passiva der Rationalisierung," *RKW-Nachrichten* 5 (1931): 128.

38. *RKW-Nachrichten* 4 (1930): 123; Feldenkirchen, *Siemens,* nn. 589–90; Brady, *Rationalization Movement;* Campbell, *Joy in Work,* 151–57.

39. "Die deutsche Rationalisierungsbewegung im Reichsmuseum—Düsseldorf," catalog and photographs of exhibit, Bundesarchiv Berlin, R 107/1, 70 (hereafter BAB). There is little available for the RKW before 1933 (R 107/2, "Reichskuratorium für Wirtschaftlichkeit, Organisation und Verwaltung," contains materials for the period 1933–39).

40. Exhibit photographs, "Die deutsche Rationalisierungsbewegung im Reichsmuseum," BAB R 107/1; Nolan, *Visions of Modernity,* 179–205. Housewives, although they did not enter the wage economy, could also partake of rationalization through an RKW program aimed at encouraging efficient housework; the catalog suggested that visitors stop by a concurrent display on the rationalization of household economy. See Helgard Kramer,

"'Rationelle Haushaltsführung' und die 'neue Frau' in der zwanziger Jahre," *Feministische Studien* 1 (1982): 123–26; and Charlotte Mühsam-Werther, "Rationalisierungsbestrebungen in der deutschen Hauswirtschaft," *Actions of the Third International Congress on the Scientific Organization of Work* (Rome, 1927), 2:345–49.

41. Harry Kessler, *In the Twenties: The Diaries of Harry Kessler*, trans. Charles Kessler (New York: Holt, Rinehart, and Winston, 1971), 399–400, 416, 412, 423.

42. Fischer, *Nazi Germany*, 243–44; Gordon A. Craig, *Germany, 1866–1945* (New York: Oxford University Press, 1981), 561.

43. Nolan, *Visions of Modernity*, 169–70; Hong, *Welfare, Modernity, and the Weimar State*, 206; Alex Zubas, "Lazy, Apathetic, and Dangerous: The Social Construction of Unemployed Workers in Germany during the Late Weimar Republic," *Contemporary European History* 10 (2001): 25–49, 26–29. See also Peter Hinrichs, *Um die Seele des Arbeiters: Arbeitspsychologie, Industrie-u. Betriebssoziologie in Deutschland, 1871–1945* (Cologne: Pahl-Rugenstein, 1981).

44. Rosenhaft, *Beating the Fascists?* 192–93, 206–7.

45. Richard J. Evans, *The Coming of the Third Reich* (New York: Penguin, 2004), 74, 272–73; "Der menschliche Faktor in der Wirtschaft," *RKW-Nachrichten* 3 (1929): 119.

46. "Wanderausstellung 'Arbeitssitz und Arbeitstisch,'" *RKW-Nachrichten* 5 (1931): 39, 74–75, 106; Hermann Gerbis, "Wanderausstellung 'Arbeitssitz und Arbeitstisch,'" *Zentralblatt für Gewerbehygiene und Unfallverhütung* 18 (April 1931): 93–95; "Ausstellung 'Arbeitssitz und Arbeitsplatz,'" *RKW-Nachrichten* 3 (1929): 155–60; "Die Sonderausstellung 'Arbeitssitz und Arbeitstisch' im Deutschen Arbeitsschutz-Museum," *Arbeitsschutz* 3 (1929): 153–56. The exhibit was part of an international tradition of exhibits on worker safety and health; more than a decade earlier Frank Gilbreth and Lillian Gilbreth had opened a fatigue museum in Providence, Rhode Island, principally displaying advances in the design of workers' chairs; see Frank B. Gilbreth and Lillian M. Gilbreth, *Fatigue Study: The Elimination of Humanity's Greatest Unnecessary Waste* (New York: Sturgis and Walton, 1916), 99–113; see also Stefan Poser, *Museum der Gefahren, Die gesellschaftliche Bedeutung der Sicherheitstechnik: Das Beispiel der Hygiene-Ausstellungen und Museum für Arbeitsschutz in Weim, Berlin und Dresden* (Münster: Waxmann, 1998).

47. "'Sonderausstellung, Arbeitssitz und Arbeitstisch,'" *Zentralblatt für Gewerbehygiene* 16 (1929): 129; "Sonderausstellung, 'Arbeitssitz und Arbeitstisch,'" *RKW-Nachrichten* 3 (1929): 119.

48. "Sonderausstellung, 'Arbeitssitz und Arbeitstisch,'" *Zentralblatt für Gewerbehygiene*, 129–30; C. W. Drescher, "Arbeitssitz und Arbeitstisch, Ein Beitrag zur Entwicklung," *Siemens Jahrbuch 1930* (Berlin: Siemens, 1930), 435–52; Drescher, "Arbeitssitz und Arbeitsplatz, Entwicklungsbeispiele aus der Praxis der Siemensbetriebe," *Arbeitsschutz*, June 15, 1929, 159. The plaques in the traveling exhibit were designed by artist Hans Möller as illustrations of Hebestreit's findings (Hermann Gerbis, "Wanderausstellung 'Arbeitssitz und Arbeitstisch,'" 94).

49. "Der Mensch und die Rationalisierung: Zweckmäßige Arbeitshaltung," *RKW-Nachrichten* 5 (1931): 124.

50. Ibid., 124, 123.

51. Drescher, "Arbeitssitz und Arbeitsplatz," 159–75.

52. Ibid., 159, 164, 166, 170.

53. "Sonderausstelling 'Arbeitssitz und Arbeitstisch,'" *Zentralblatt für Gewerbehygiene,* 129.

54. Harry Braverman, *Labor and Monopoly Capital: The Degradation of Work in the Twentieth Century* (New York: Monthly Review Press, 1974), 125; Theodore M. Porter, *Trust in Numbers: The Pursuit of Objectivity in Science and Public Life* (Princeton: Princeton University Press, 1995), 89; Steven Lubar, "Representation and Power," *Technology and Culture* 36 (1995): S55.

55. In parallel to Max Weber's conception of discipline as a way to guarantee that military orders would be obeyed; see Philip S. Gorski, *The Disciplinary Revolution: Calvinism and the Rise of the State in Early Modern Europe* (Chicago: University of Chicago Press, 2003), 32; and Poser, *Museum der Gefahren,* 224–28.

56. Brady, *Rationalization Movement,* 403.

57. Ibid., xx, 21–22.

58. Military drill was one of Michel Foucault's examples of discipline, resulting in a dressage of human gestures that could also be seen in how prisoners moved. See Foucault, *Discipline and Punish: The Birth of the Prison* (New York: Random House, 1977).

59. Feldenkirchen, *Siemens,* 185–92.

60. Hebestreit, "Physiologische Grundlagen," 12.

61. "Rationalisierung als Bewegung," *RKW-Nachrichten* 4 (1930): 121–23; "Berufsauslese, Berufsausbildung und Bestgestaltung der Arbeit," *RKW-Nachrichten* 5 (1931): 12–13.

62. Feldenkirchen, 345, 661 n. 1; Sachse, *Siemens,* 30–31.

63. Nolan, *Visions of Modernity,* 155, 178, 169, 167; Hinrichs, *Um die Seele des Arbeiters,* 303–4; Homburg, *Industriearbeit und Rationalisierung,* 677–78; see also Heidrun Homburg, "The 'Human Factor' and the Limits of Rationalization: Personnel Management Strategies and the Rationalization Movement in German Industry between the Wars," in *The Power to Manage? Employers and Industrial Relations in Comparative Historical Perspective,* ed. Steven Tolliday and Jonathan Zeitlin (London: Routledge, 1991), 147–75.

64. Andrew Pickering, "The History of Economics and the History of Agency," in *The State of the History of Economics: Proceedings of the History of Economics Society,* ed. James P. Henderson (London: Routledge, 1997), 6–18, 9; James C. Scott, *Seeing like a State: How Certain Schemes to Improve the Human Condition Have Failed* (New Haven, CT: Yale University Press, 1998), 349.

Six • *Pride in Efficiency*

1. Robert William Fogel and Stanley L. Engerman, *Time on the Cross: The Economics of American Negro Slavery* (New York: Little, Brown, 1974; rpt., New York: Norton, 1995). The 1989 Norton edition includes a short description of the evolution of Fogel's thought in an "Afterword 1989" (265–75), to be distinguished from "Afterword: The Moral Problem of Slavery" in Fogel's later volume *Without Consent or Contract: The Rise and Fall of American Slavery* (New York: Norton, 1989).

2. Fogel, *Without Consent or Contract,* 410–11, 10.

3. Ibid., 411.

4. Robert William Fogel, "The Long Reach of the Ideological Issues That Produced the Civil War," in *Without Consent or Contract: The Rise and Fall of American Slavery: Evidence and Methods,* ed. Robert William Fogel, Ralph A. Galantine, and Richard L. Manning (New York: Norton, 1992), 46.

5. Alfred H. Conrad and John R. Meyer, "The Economics of Slavery in the Ante Bellum South," *Journal of Political Economy* 66 (1958): 95–130. Quotation from *Time on the Cross,* 67.

6. Kenneth M. Stampp, "Introduction: A Humanistic Perspective," in *Reckoning with Slavery: A Critical Study in the Quantitative History of American Negro Slavery,* ed. Paul A. David et al. (New York: Oxford University Press, 1976), 1–30, 12–13.

7. Robert William Fogel, *The Slavery Debates, 1952–1990: A Retrospective* (Baton Rouge: Louisiana State University Press, 2003), 21–22, 20.

8. See Stampp, "Introduction," 2–3; and Peter Novick, *That Noble Dream: The "Objectivity" Question and the American Historical Profession* (Cambridge: Cambridge University Press, 1988), 384.

9. Fogel, *Without Consent or Contract,* 390–91.

10. Conrad and Meyer, "Economics of Slavery," 121–22, 115.

11. Fogel, *Without Consent or Contract,* 391.

12. Stanley M. Elkins, "The Slavery Debate," *Commentary* (December 1974): 40. Elkins cited an as yet unpublished paper by R. M. Hartwell, "Slave Labor and Factory Labor."

13. Robert Coles, *Farewell to the South* (Boston: Little, Brown, 1972).

14. Robert William Fogel and Stanley L. Engerman, "Table B.1: Major Bodies of Data Collected by Cliometricians," *Time on the Cross: Evidence and Methods—A Supplement* (Boston: Little, Brown, 1974), 22–25.

15. Fogel and Engerman, *Time on the Cross* (1995 ed.), 259.

16. Ibid., 228–32.

17. Fogel and Engerman, *Time on the Cross: Evidence and Methods,* 221–22.

18. Peter Passell, *New York Times Book Review,* April 28, 1974, 4.

19. Fogel and Engerman, *Time on the Cross* (1995 ed.), 4–6.

20. Soma Golden, "New Slavery Book Kindles a Dispute," *New York Times,* May 2, 1974, 49, 57; Passell, *New York Times Book Review,* 4; Walter Clemons, "A New Look at Slavery," *Newsweek,* May 6, 1974, 77–78; "Massa's in de Cold, Cold Computer," *Time,* June 17, 1974, 98–100; Thomas Fleming, "The 'Real' Uncle Tom," *Reader's Digest,* March 1975, 124–28.

21. David Brion Davis, "Slavery and the Post–World War II Historians," *Daedalus* (Spring 1974): 9–10.

22. "Massa's in de Cold, Cold Computer," 99.

23. Clemons, "New Look at Slavery," 78.

24. Winthrop Jordan, "The Tone of Snake-Root Salesmen," *Psychology Today* 8 (September 1974): 119, 122.

25. Fleming, "'Real' Uncle Tom," 124–28; quoted in Stampp, "Introduction," 7.

26. M. E. Bradford (340–42, 359, 341) and Jeffrey Hart (340–42, 359), both in *National Review*, March 28, 1975.

27. Christopher Dell, "In Money Terms, It Worked Fine," *Nation* 219, October 5, 1974, 310–11.

28. Stampp, "Introduction," 8.

29. Thomas J. Haskell, "The True and Tragical History of *Time on the Cross*," *Objectivity Is Not Neutrality: Explanatory Schemes in History* (Baltimore: Johns Hopkins University Press, 1998), 47; originally published in *New York Review of Books*, October 2, 1975, 33–39.

30. Haskell, "True and Tragical History," 46, 48–49, 50–51, 53, 52, 56.

31. Peter Kolchin, "More *Time on the Cross?* An Evaluation of Robert William Fogel's *Without Consent or Contract*," *Journal of Southern History* 58 (1992): 492.

32. Bridenbaugh, quoted in Novick, *That Noble Dream*, 384.

33. Haskell, "True and Tragical History," 54, 53.

34. Eugene D. Genovese, *The Political Economy of Slavery: Studies in the Economy and Society of the Slave South* (New York: Random House, 1965), 46, 50.

35. Fogel and Engerman, *Time on the Cross: Evidence and Methods*, 126–31.

36. Paul David and Peter Temin, "Slavery: The Progressive Institution?" in David et al., *Reckoning with Slavery*, 202–3; originally published in *Journal of Economic History* 34 (September 1974): 739–83.

37. Ibid., 225–26.

38. The critical volumes included Herbert Gutman, *Slavery and the Numbers Game: A Critique of* Time on the Cross (Urbana: University of Illinois Press, 1975); Gar Walton, ed., "A Symposium on *Time on the Cross*," *Explorations in Economic History* 12 (1975); and David et al., *Reckoning with Slavery*. The *American Economic Review* exchange included Fogel and Engerman, "Explaining the Relative Efficiency of Slave Agriculture in the Antebellum South," *American Economic Review* 67 (1977): 275–96; Paul A. David and Peter Temin, "Explaining the Relative Efficiency of Slave Agriculture in the Antebellum South: Comment," *American Economic Review* 69 (1979): 213–18; Gavin Wright, "The Efficiency of Slavery: Another Interpretation," *American Economic Review* 69 (1979): 219–26; D. F. Schaefer and M. D. Schmitz, "The Relative Efficiency of Slave Agriculture: A Comment," *American Economic Review* 69 (1979): 208–12; Thomas Haskell, "Explaining the Relative Efficiency of Slave Agriculture in the Antebellum South: A Reply to Fogel and Engerman," *American Economic Review* 69 (1979): 206–7; and Fogel and Engerman, "Explaining the Relative Efficiency of Slave Agriculture in the Antebellum South: Reply," *American Economic Review* 70 (1980): 672–90. Also see Roger L. Ransom, *Conflict and Compromise: The Political Economy of Slavery, Emancipation, and the American Civil War* (Cambridge: Cambridge University Press, 1989); Michael Tadman, *Speculators and Slaves: Masters, Traders, and Slaves in the Old South* (Madison: University of Wisconsin Press, 1989).

39. Fogel, *Slavery Debates*, 32.

40. Ibid., 27, 33.

41. Fogel and Engerman, *Time on the Cross*, 203–4.

42. Ibid., 202.

43. Ibid., 203.

44. Ibid., 205.

45. Ibid., 210, 202–4, 208.

46. Ibid., 213.

47. Ira Berlin, *Generations of Captivity: A History of African-American Slaves* (Cambridge, MA: Belknap Press, 2003), 64.

48. Fogel, *Without Consent or Contract*, 190; from Sidney W. Mintz, "Was the Plantation Slave a Proletarian?" *Review* 2 (1978): 93, 94–96.

49. Herbert G. Gutman and Richard Sutch, "Sambo Makes Good, or Were Slaves Imbued with the Protestant Work Ethic?" in *Reckoning with Slavery*, 68.

50. Fogel, *Without Consent or Contract*, 26.

51. Fogel, *Without Consent or Contract*, 27; Fogel and Engerman, *Time on the Cross*, 204.

52. Fogel and Engerman, *Time on the Cross*, 237, 232.

53. Gutman and Sutch, "Sambo Makes Good," 65–67; Kenneth M. Stampp, *The Peculiar Institution: Slavery in the Ante-Bellum South* (New York: Knopf, 1956), 186.

54. Gutman and Sutch, "Sambo Makes Good," 93.

55. Fogel, *Without Consent or Contract*, 411, 394.

56. Michel Foucault, "The Ethics of the Concern of the Self as a Practice of Freedom," interview with H. Becker, R. Fornet-Betancourt, and A. Gomez-Müller, January 20, 1984, *Michel Foucault: Ethics, Subjectivity and Truth* (New York: New Press, 1994), 1:292; Michel Foucault, "Structuralism and Post-Structuralism," interview with Gérard Raulet, 1983, *Michel Foucault:. Aesthetics, Method, and Epistemology* (New York: New Press, 1994), 2:450–53; Catherine A. Foote and Arthur W. Frank, "Foucault and Therapy: The Disciplining of Grief," in *Reading Foucault for Social Work,* ed. Adrienne S. Chambon, Allan Irving, and Laura Epstein (New York: Columbia University Press, 1999), 172–73.

57. Robert William Fogel, "A Note on the Effect of Moral Issues on the Debates among Cliometricians," in Fogel, Galantine, and Manning, *Without Consent or Contract: Evidence and Methods,* 591.

58. Kolchin, "More *Time on the Cross?"* 498.

59. Harold D. Woodman, "Capitalism, Morality, and the Peculiar Institution," *Journal of Economic History* 55 (1995): 370.

60. Ralph A. Austen, review of *Without Consent or Contract, Business History Review* 64 (1990): 776.

61. David Brion Davis, *Challenging the Boundaries of Slavery* (Cambridge: Harvard University Press, 2003), 31–32, 57.

62. Fogel, *Without Consent or Contract*, 410.

63. Fogel, *Slavery Debates*, 68.

Seven • Global Efficiency

1. Stephen Kern, *The Culture of Time and Space, 1880–1918* (Cambridge, MA: Harvard University Press, 1983), 92.

2. Joseph E. Stiglitz, *Globalization and Its Discontents* (New York: Norton, 2003), ix; Steven Best and Douglas Kellner, *The Postmodern Adventure: Science, Technology, and Cultural Studies at the Third Millennium* (New York: Guilford Press, 2001), 207.

3. Krishan Kumar, *From Post-Industrial to Post-Modern Society: New Theories of the Contemporary World*, 2nd ed. (Malden, MA: Blackwell, 2005), 7.

4. Daniel Bell, *The Coming of Post-Industrial Society: A Venture in Social Forecasting* (New York: Basic Books, 1973); Margaret A. Rose, *The Post-Modern and the Post-Industrial: A Critical Analysis* (Cambridge: Cambridge University Press, 1991), 21–39; Kumar, *From Post-Industrial to Post-Modern Society*, 29–32.

5. Stiglitz, *Globalization and Its Discontents*, 218, 73.

6. Janet Abbate, *Inventing the Internet* (Cambridge, MA: MIT Press, 1999); Paul N. Edwards, *The Closed World: Computers and the Politics of Discourse in Cold War America* (Cambridge, MA: MIT Press, 1996).

7. Adam Liptak, "Filing a Heartfelt Appeal against the Legal System," *New York Times*, May 15, 2004, A19.

8. Thane Rosenbaum, *The Myth of Moral Justice: Why Our Legal System Fails to Do What's Right* (New York: HarperCollins, 2004), 20.

9. *Connick v. Myers*, 461 US, 146. Justice White wrote the opinion, joined by Chief Justice Burger and justices Powell, Rehnquist, and O'Connor; justices Brennan, Marshall, Blackmun, and Stevens dissented, with Brennan writing the dissenting opinion.

10. *Pickering v. Board of Education of Township High School District 205, Will County, No. 510*, 391 US, 571–72.

11. Jeffrey S. Strauss, "Dangerous Thoughts? Academic Freedom, Free Speech, and Censorship Revisited in a Post–September 11th America," *Washington University Journal of Law and Policy* 15 (2004): 343–67.

12. *Pickering v. Board of Education*, 563–84.

13. Ibid., 568.

14. Justice White, in his partial dissent, agreed that Pickering's letter did not display a knowing or reckless disregard for the truth but thought that was enough to settle the matter, without raising questions about the harm it might have done.

15. A classic statement of this dilemma is Arthur M. Okun, *Equality and Efficiency: The Big Tradeoff* (Washington, DC: Brookings Institution, 1975).

16. James J. Heckman, "Investing in Disadvantaged Young Children Is an Economically Efficient Policy," paper presented January 10, 2006, www.ced.org/projects/prek-shtml #2006conference, accessed August 14, 2006.

17. D. T. Ellwood, "The Sputtering Labor Force of the Twenty-first Century: Can Social Policy Help?" in *The Roaring Nineties: Can Full Employment Be Sustained?* ed. A. Krueger and R. Solow (New York: Russell Sage Foundation, 2001), 421–89; cited in Heckman, "Investing in Disadvantaged Young Children," 9.

18. Charles S. Maier, "The Factory as Society: Ideologies of Management in the Twentieth Century," in *Ideas into Politics: Aspects of European History, 1880 to 1950*, ed. R. J. Bullen, H. Pogge von Strandmann, and A. B. Polonsky (London: Croom Helm, 1974), 147–63, 155–59.

19. See Raymond E. Callahan, *Education and the Cult of Efficiency: A Study of the Social Forces That Have Shaped the Administration of the Public Schools* (Chicago: University of Chicago Press, 1962); Char Roone Miller, *Taylored Citizenship: State Institutions and Subjectivity* (Westport, CT: Praeger, 2002).

20. The conference was cohosted by the McKnight Foundation, with the cooperation of the University of Minnesota.

21. "Ready to Learn," *Minneapolis–St. Paul StarTribune*, December 15, 2003, 24A.

22. Art Rolnick and Rob Grunewald, "Early Childhood Development: Economic Development with a High Public Return," *The Region: Banking and Policy Issues Magazine* (December 2003): 6–12.

23. Art Rolnick and Rob Grunewald, "Early Childhood Development: Economic Development with a High Public Return," *FedGazette*, Federal Reserve Bank of Minneapolis, January 24, 2003, 6–11; Lawrence J. Schweinhart, Helen V. Barnes, and David P. Weikart, *Significant Benefits: The High/Scope Perry Preschool Study through Age 27* (Ypsilanti, MI: High/Scope Press, 1993), 32.

24. Rolnick and Grunewald, "Early Childhood Development," 10, table 1B.

25. Jean Burr and Rob Grunewald, "Lessons Learned: A Review of Early Childhood Development Studies," technical paper, April 2006, iii, 7, www.minneapolisfed.org/research/studies/earlychild.

26. Rolnick and Grunewald, "Early Childhood Development," 11. The internal rate of return describes the interest rate earned on a business investment, in which payments and revenue are figured at regular intervals. See also Rob Grunewald and Art Rolnick, "Early Childhood Development on a Large Scale," *The Region: Banking and Policy Issues Magazine*, June 2005, www.minneapolisfed.org/pubs/region/05-06/ecd.cfm; Rob Grunewald and Art Rolnick, "A Proposal for Achieving High Returns on Early Childhood Development," presented December 3, 2004, in Washington, DC, at the conference "Building the Economic Case for Investments in Preschool," www.minneapolisfed.org/research/studies/earlychild.

27. Janet Currie, "Getting Inside the 'Black Box' of Head Start Quality: What Matters and What Doesn't," and James Heckman, "Interpreting the Evidence of Family Influence on Child Development," both from "The Economics of Early Childhood Development: Lessons for Economic Policy" conference, Federal Reserve Bank of Minneapolis, October 17, 2003; both Currie and Heckman quoted in Kathy Cobb, "The ABCs of ECD," *Region: Banking and Policy Issues Magazine* (December 2003): 2.

28. Rip Rapson, "Opening Remarks," "The Economics of Early Childhood Development: Lessons for Economic Policy" conference, Federal Reserve Bank of Minneapolis, October 17, 2003, 2.

29. Neal St. Anthony, "Investing in Kids Pays Off, Execs Agree," *StarTribune*, March 13, 2004, 1D.

30. Lori Sturdevant, "Business Shouldn't Delay Going to Bat for Young Learners," *StarTribune*, March 21, 2004, 2AA.

31. "Letters from Readers," *StarTribune*, December 19, 2003, 32A.

32. Ibid., December 17, 2003, 20A.

33. Gary Rivlin, "Who's Afraid of China?" *New York Times*, December 19, 2004, B1; Thomas L. Friedman, *The World Is Flat: A Brief History of the Twenty-First Century* (New York: Farrar, Straus and Giroux, 2005), 414–25.

34. David F. Noble, "Social Change in Machine Design: The Case of Automatically Controlled Machine Tools, and a Challenge for Labor," *Politics and Society* 8 (1978): 313–47.

35. Shoshana Zuboff, *In the Age of the Smart Machine: The Future of Work and Power* (New York: Basic Books, 1988), 346, 330, 355.

36. James W. Cortada, *The Digital Hand: How Computers Changed the Work of American Manufacturing, Transportation, and Retail Industries* (New York: Oxford University Press, 2004), 361–63, 118–20, 212–13.

37. Ibid., 318–36, 161–78, 177.

38. Manuel Castells, *The Rise of the Network Society: The Information Age: Economy, Society, and Culture,* 2nd ed. (Malden, MA: Blackwell, 2000), 164–79, 291–302.

39. Ibid., 291–302, 170.

Conclusion: The Future of Efficiency

1. D. S. L. Cardwell, *From Watt to Clausius: The Rise of Thermodynamics in the Early Industrial Age* (Ithaca, NY: Cornell University Press, 1971), 211–12.

2. Quoted in ibid., 192.

3. There have been a number of theories of control, ranging from the early modern use of a "contre-rolle," or duplicate record to verify accounts, and of double-entry bookkeeping to twentieth-century theories of feedback and control in cybernetics and the sociological concept of social control. Differences between these theories turn largely on the methods or processes by which control is achieved. See Miriam R. Levin, "Contexts of Control," in *Cultures of Control,* ed. Miriam R. Levin (Amsterdam: Harwood Academic Publishers, 2000), 13–40; Mary Poovey, *A History of the Modern Fact: Problems of Knowledge in the Sciences of Wealth and Society* (Chicago: University of Chicago Press, 1998), 33–65; David A. Mindell, *Between Human and Machine: Feedback, Control, and Computing before Cybernetics* (Baltimore: Johns Hopkins University Press, 2002), 2–11.

4. Sadi Carnot, *Reflections on the Motive Force of Fire: A Critical Edition with the Surviving Scientific Manuscripts,* ed. and trans. Robert Fox (Manchester: Manchester University Press, 1986), 61–62, 111–13.

5. Eric D. Weitz, *A Century of Genocide: Utopias of Race and Nation* (Princeton: Princeton University Press, 2003), 102–43.

6. On Birkenau, see note 7, p. 171, esp. Allen, "Modernity, the Holocaust, and Machines without History." Dan Diner, *Beyond the Conceivable: Studies on Germany, Nazism, and the Holocaust* (Berkeley: University of California Press, 2000), 148. Diner concludes that the Nazi system was nonrational because it valued destruction over labor.

7. Langdon Winner, "Do Artifacts Have Politics?" *Daedalus* 109 (1980): 127.

8. Mark C. Taylor, *The Moment of Complexity: Emerging Network Culture* (Chicago: University of Chicago Press, 2001), esp. 28–30; Stephen Toulmin, *Cosmopolis: The Hidden Agenda of Modernity* (Chicago: University of Chicago Press, 1990), 198–201.

9. Toulmin, *Cosmopolis,* 180–86.

10. Romano Guardini, *The End of the Modern World* (Wilmington, DE: ISI Books, 1998), 193.

The literature on efficiency is almost entirely devoted to studies of its use in particular contexts; this literature is discussed here in the entry for the appropriate chapter. Two themes are especially important: the relationship of efficiency to modernity and the role of efficiency as a tool of human agency. An important point of overlap is the transition in early modernity from an emphasis on divine causal agency to an emphasis on human technological agency.

The literature on modernity and technology is enormous. Essential is the collection edited by Thomas J. Misa, Philip Brey, and Andrew Feenberg, *Modernity and Technology* (Cambridge, MA: MIT Press, 2003). The editors survey and critique treatments of modernity in the lead articles, and the contributed papers make a sustained effort to bring detailed analysis of technology into conversation with the deep but often abstract theoretical literature. The foundational relationship for modernity between concepts of human emancipation and enlightened rationality is little treated in the literature on technology; that precise junction is explored in depth by Alain Touraine in *Critique de la modernité* (Paris: Fayard, 1991). Touraine argues that the foundations of rationalization and means-end rationality are to be found in this relationship, although he does not treat technologies with any specificity. Nor does Max Weber, the figure most closely associated with conceptions of rationalization, but central to his analysis is the question of effective human agency and theological belief; see Weber, *The Protestant Ethic and the Spirit of Capitalism* (London: Routledge, 1930). Zygmunt Bauman takes up the question of emancipation, freedom, and human agency in *Liquid Modernity* (Cambridge: Polity Press, 2000). David F. Noble has demonstrated the increasing association of technology with human rather than divine power; see *The Religion of Technology: The Divinity of Man and the Spirit of Invention* (New York: Knopf, 1998). The ascendancy of stories of human agency through technology is analyzed for the American context by David Nye in *America as Second Creation: Technology and Narratives of New Beginnings* (Cambridge, MA: MIT Press, 2003). Thomas Parke Hughes suggests a similar theme in his survey of technology in history, in *Human-Built World: How to Think about Technology and Culture* (Chicago: University of Chicago Press, 2004).

The importance of the shift in attributing creative agency from deity to human is discussed by David H. Hopper in *Technology, Theology, and the Idea of Progress* (Louisville:

Westminster / John Knox Press, 1991). A similar issue is taken up by Don Ian Smith in "Forest Management and the Theology of Nature," *American Forests* 99 (November–December 1993): 13–16.

The idea of a machine was itself closely linked with theological conceptions of effective agency in medieval times; see Günter Ropohl, "Die Maschinenmetapher," *Technikgeschichte* 58 (1991): 3–14. Technology is rarely treated in detail in discussions of its relationship to theology, but Carl Mitcham set an ambitious framework for the examination in his edited collection, *Theology and Technology: Essays in Christian Analysis and Exegesis* (Lanham, MD: University Press of America, 1984). The most sustained treatment of efficiency is the dense and often frustrating classic by the French lay theologian Jacques Ellul, *The Technological Society* (New York: Knopf, 1964). It continues to repay close reading, and I have become less convinced that his may be dismissed as a deterministic analysis.

Efficiency is allied with people's visions of how the world does, and should, work. Jim Corkery examines the relationship between visions and technology in "Does Technology Squeeze Out Transcendence—or What?" in *Technology and Transcendence,* ed. Michael Breen, Eamonn Conway, and Barry McMillan (Dublin: Columba Press, 2003). For a consideration of the conflict between visions of order and voices of hope, see Walter Brueggemann, "Living toward a Vision: Grief in the Midst of Technique," *Hope within History* (Atlanta: John Knox Press, 1987), 72–91. The relationship between technology, modernity, and intellectual visions of society and the future is the subject of Mikael Hård and Andrew Jamison's edited collection *The Intellectual Appropriation of Technology: Discourses on Modernity, 1900–1939* (Cambridge, MA: MIT Press, 1998). The importance of human initiative in making visions into reality is discussed in Andrew Feenberg, *Transforming Technology: A Critical Theory Revisited* (Oxford: Oxford University Press, 2002). For a striking study of the collision between visions of modernity and the limits of human agency, see Manuel A. Vásquez, *The Brazilian Popular Church and the Crisis of Modernity* (Cambridge: Cambridge University Press, 1998).

Modern efficiency was an accounting technique; there are several interesting treatments of the development of accounting as a method of management and control. Miriam Levin locates the history of the concept of control in the use of duplicate, counter records in early modern accounting; see Levin, "Contexts of Control," *Cultures of Control* (Amsterdam: Harwood Academic Publishers, 2000), 13–39. A more detailed analysis appears in Mary Poovey's study of record keeping and numerical representation, in *A History of the Modern Fact: Problems of Knowledge in the Sciences of Wealth and Society* (Chicago: University of Chicago Press, 1998). The authority of numbers and measurements, including measures of efficiency, is analyzed by Theodore M. Porter in *Trust in Numbers: The Pursuit of Objectivity in Science and Public Life* (Princeton: Princeton University Press, 1995); and in the collection *The Values of Precision,* ed. Norman M. Wise (Princeton: Princeton University Press, 1995).

Static and Dynamic Efficiency

John Smeaton is a central figure in the history of engineering, and his measurements of the output and input of waterwheels are an example of this type of quantification from relatively early in industrialization. The Franklin Institute's trials illustrate another way in which such measurements were made. The main body of materials analyzed in this chapter consists of Smeaton's reports on his experimental work quantifying machine performance generally and the Franklin Institute's reports on its own experiments in quantifying waterwheel performance. Smeaton's scientific reports appeared in the *Philosophical Transactions of the Royal Society* between 1759 and 1782; the Franklin Institute reported its progress and findings in its own journal, the *Journal of the Franklin Institute*, from 1831 to 1841. The Royal Society in London has a collection of Smeaton's drawings (which will be described later) as well as minutes on the awarding of the Copley Medal (Royal Society Council Minutes, 1747–63, November 22, 1759, 4:221–22). Smeaton corresponded with instrument maker Benjamin Wilson about his scientific interests; that correspondence is available in the British Library (British Library Additional Manuscripts, no. 30094, John Smeaton to Benjamin Wilson, letters, 1744–64). The Institution of Civil Engineers Archives, London, has a private letter book from 1764, machine letter books from 1781–92, and notes of accounts in Smeaton's workshop in 1751–52. Other materials are described in the essential volume edited by A. W. Skempton, *John Smeaton, FRS* (London: Thomas Telford, 1981). It contains a biographical study by John Trevor, an analysis of Smeaton's scientific work by Norman Smith, and a description of Smeaton's professional practice by Denis Smith; it also contains a list of Smeaton's construction works.

On any question of waterwheels the place to begin is Terry S. Reynolds, *Stronger Than a Hundred Men: A History of the Vertical Waterwheel* (Baltimore: Johns Hopkins University Press, 1983); Reynolds is especially good on the history of attempts to analyze how a waterwheel works and presents complicated and highly abstract concepts clearly but without undue simplification. Paul N. Wilson analyzed Smeaton's surviving waterwheel drawings in "The Waterwheels of John Smeaton," *Transaction of the Newcomen Society* 30 (1955): 25–48. The first volume of the first edition of Samuel Smiles's *Lives of the Engineers* (London: John Murray, 1861) is largely devoted to Smeaton. A recent mathematical analysis is M. Denny, "The Efficiency of Overshot and Undershot Waterwheels," *European Journal of Physics* 25 (2004): 193–202. Smeaton's role in the developing engineering profession is discussed in R. A. Buchanan, *The Engineers: A History of the Engineering Profession in Britain, 1750–1914* (London: Jessica Kingsley, 1989).

Smeaton's waterwheels study was reprinted widely, for example in Oliver Evans, *The Young Mill-Wright and Miller's Guide* (Philadelphia: Blanchard and Lea, 1860), and also translated into French; see Smeaton, *Recherches expérimentales sur l'eau et le vent*, trans. M. P. S. Girard (1810; rpt., Paris: Courcier/Bachelier, 1827). His papers and reports were collected and printed in the decades following his death in 1792; A. W. Skempton describes them in "Papers, Reports and Drawings," in *John Smeaton, FRS*. The Royal Society has in its possession a collection of Smeaton's engineering drawings, many annotated by the engineer John Farey; see H. W. Dickinson and A. A. Gomme (for the Newcomen Society), *A Catalogue of the Civil and Mechanical Engineering Designs, 1741–1792, of John Smeaton* (London: Printed by the Courier Press, 1950); and A. P. Woolrich, "John Farey and the Smeaton Manuscripts," *History of Technology* 10 (1985): 181–216. Smeaton is a frequent subject of notices and reports in the *Transactions of the Newcomen Society*.

Smeaton's role as a mediator between scientific and engineering contexts has been described by Margaret Jacob in three studies of the relationship between the scientific revolution and industrialization: *The Cultural Meaning of the Scientific Revolution* (New York: McGraw Hill, 1988); *Scientific Culture and the Making of the Industrial West* (New York: Oxford University Press, 1997); and *Practical Matter: Newton's Science in the Service of Industry and Empire, 1867–1851* (Cambridge: Harvard University Press, 2004), coauthored with Larry Stewart. Stewart's study of the role of natural philosophical knowledge in early industrialization provides an important context to Smeaton's work; see *The Rise of Public Science: Rhetoric, Technology, and Natural Philosophy in Newtonian Britain, 1660–1750* (Cambridge: Cambridge University Press, 1992). Stewart calls on historians to investigate more thoroughly the relationship between science and technology during industrialization, a relationship of which Smeaton is a prime example, drawing on the classic study of A. E. Musson and Eric Robinson, *Science and Technology in the Industrial Revolution* (Manchester: Manchester University Press, 1969). Other explorations of this relationship are by Ken Alder, "French Engineers Become Professionals; or, How Meritocracy Made Knowledge Objective," in *The Sciences in Enlightened Europe*, ed. William Clark, Jan Golinski, and Simon Schaffer (Chicago: University of Chicago Press, 1999), 94–125; Larry Stewart and Paul Weindling, "Philosophical Threads: Natural Philosophy and Public Experiment among the Weavers of Spitalfields," *British Journal for the History of Science* 28 (1995): 37–62; and a pair of articles by Simon Schaffer, "Machine Philosophy: Demonstration Devices in Georgian Mechanics," *Osiris* ser 2, 9 (1994): 157–82, and "The Show That Never Ends: Perpetual Motion in the Early Eighteenth Century," *British Journal for the History of Science* 28 (1998): 157–89. Jenny Uglow's recent study *The Lunar Men: Five Friends Whose Curiosity Changed the World* (New York: Farrar, Straus and Giroux, 2002) also develops this theme; Robert Schofield's study *The Lunar Society of Birmingham: A Social History of Provincial Science and Industry in Eighteenth-Century England* (Oxford: Clarendon Press, 1963) remains a classic.

The role of philosophical debates about motion and its quantification is addressed for the practical American context by Edwin T. Layton Jr. in "Newton Confronts the American Millwrights, or, Action and Reaction Are Not Always Equal," in *Beyond History of Science*, ed. Elizabeth Garber (Bethlehem: Lehigh University Press, 1990), 179–93. Mary Terrall has returned to the debate, in the French context, in "*Vis viva* Revisited," *History of Science* 42 (2004): 189–209. The question interested historians from the 1960s into the 1970s. In "Eighteenth Century Attempts to Solve the *Vis Viva* Controversy," *Isis* 56 (1965): 281–97, Thomas L. Hankins described how difficult to comprehend were the concepts involved; and L. L. Laudan summarized the historical research in "The *Vis Viva* Controversy: A Post-Mortem," *Isis* 59 (1968): 131–43. Carolyn Iltis analyzed Leibniz's conception in "Leibniz and the *Vis Viva* Controversy," *Isis* 62 (1971): 21–35; and David Papineau asked how significant were the distinctions in dispute in "The *Vis Viva* Controversy: Do Meanings Matter?" *Studies in the History and Philosophy of Science* 8 (1977): 111–37. Wilson L. Scott considered Smeaton one of three English engineers, along with William Wollaston and Peter Ewart, who "crossed the Rubicon" by defining force or power as effect over distance, rather than over time, as did Leibniz and in contrast to most of their English contemporaries; see Scott, *The Conflict between Atomism and Conservation Theory* (London: MacDonald, 1970). Smeaton's physics is analyzed by Silvio Bergia in "La Descrizione dei fenomeni meccanici in termini energetici nell'opera di John Smeaton," *Giornale di Fisica della Società Italiana di Fisica* 22 (1981): 295–310; 23 (1982): 59–73. Gerhard Zweckbronner took up the question of *vis viva* as it had been applied in mechanic's practice, with disastrous results for a waterwheel of Poncelet's erected in Württemberg, in "Das Prinzip der lebendigen Kräfte, ein Bindeglied zwischen rationeller Mechanik und Praktischem Maschinenbau im Zeitalter der Industrialisierung," *Technik Geschichte* 48 (1981): 89–111. An especially interesting analysis appears in Kevin C. de Berg's discussion of how Smeaton's work might be used in helping to give physics and engineering students a physical understanding of the concept of mechanical work, rather than just a mathematical one; see Berg, "The Development of the Concept of Work: A Case Where History Can Inform Pedagogy," *Science and Education* 6 (1997): 511–27.

An important question has been how much the mechanical arts were indeed influenced by philosophical debate. Clifford Truesdell argued that they were not ("A Program toward Recovering the Rational Mechanics of the Age of Reason," *Archive for History of Exact Science* 1 [1960]: 1–36), while Ivor Grattan-Guinness argued later that practical uses were foremost in the minds of the savants responsible for defining the concept of work ("Work for the Workers: Advances in Engineering Mechanics and Instruction in France, 1800–1830," *Annals of Science* 41 [1984]: 1–33, 7). Alan Q. Morton uses the concept of public science to analyze the mutual influences in "Concepts of Power: Natural Philosophy and the Uses of Machines in Mid-Eighteenth-Century London," *British Journal for the History of Science* 28 (1995): 63–78.

Smeaton also devoted a good part of his career working to improve steam engine performance, although more famous in this regard was James Watt, whose most important work came during the later part of Smeaton's career. Ben Marsden gives a nice overview in the short text *Watt's Perfect Engine: Steam and the Age of Invention* (New York: Columbia University Press, 2002), which includes an excellent and selective bibliography. On conceptions of effect D. S. L. Cardwell's study of the laws of energy, *From Watt to Clausius: The Rise of Thermodynamics in the Early Industrial Age* (London: Heinemann, 1971), is especially helpful, as is the essay by Robert Fox, "Watt's Expansive Principle in the Works of Sadi Carnot and Nicolas Clément," *Notes and Records of the Royal Society* 24 (1969): 233–53. The Enlightenment context, which Smeaton and Watt shared, is described by Christine MacLeod in "James Watt, Heroic Invention, and the Idea of the Industrial Revolution," in *Technological Revolutions in Europe: Historical Perspectives*, ed. Maxine Berg and Kristine Bruland (Cheltenham: Edward Elgar, 1998), 96–115; and Peter J. Jones, "Living the Enlightenment and the French Revolution: James Watt, Matthew Boulton, and Their Sons," *Historical Journal* 42 (1999): 157–82.

Bruce Sinclair's book *Philadelphia's Philosopher Mechanics: A History of the Franklin Institute, 1824–1865* (Baltimore: Johns Hopkins University Press, 1974) remains the definitive study of the Franklin Institute; he addressed the scientific work specifically in "Science, Technology, and the Franklin Institute," in *Pursuit of Knowledge in the Early American Republic: American Scientific and Learned Societies from Colonial Times to the Civil War*, ed. Alexandra Oleson and Sanborn C. Brown (Baltimore: Johns Hopkins University Press, 1974), 194–207. The *Journal of the Franklin Institute* contains the published waterwheels reports. Adrian Michal McMahon and Stephanie A. Morris examined the institute's committee for testing machine models in the introduction to *Technology in Industrial America: The Committee on Science and the Arts of the Franklin Institute, 1824–1900* (Wilmington, DE: Scholarly Resources, 1977).

An indispensable resource for early American technology is Nina A. Lerman's exhaustive bibliography, "Books on American Technology, 1966–1991," in *Early American Technology: Making and Doing Things from the Colonial Era to 1850*, ed. Judith A. McGaw (Chapel Hill: Published for the Institute of Early American History and Culture, Williamsburg, VA, by the University of North Carolina Press, 1994), 358–429. "Brooke Hindle's Pre-1966 Bibliography," an appendix in the same volume, is also very useful (431–60). The machining trades in Philadelphia are discussed in Eugene S. Ferguson, ed., *The Early Engineering Reminiscences (1815–1840) of George Escol Sellers*, United States National Museum Bulletin No. 238 (Washington, DC: Smithsonian Institution, 1965); and Anthony F. C. Wallace describes the surrounding Pennsylvania context in *Rockdale: The Growth of an American Village in the Early Industrial Revolution* (New York: Norton, 1978). Nina Lerman's own studies of gender and technical education in Philadelphia provide a necessary corrective to thinking of technology in terms of "masculine" prime movers; see "'Preparing for the Duties and Practical Business of Life': Technological Knowledge and Social Structure in Mid-19th-Century Philadelphia," *Technology and Culture* 38 (1997): 31–59; and "The Uses of Useful Knowledge: Science, Technology, and Social Boundaries in an Industrializing City," *Osiris* 12 (1997): 39–59. Two useful studies treating coming developments are Paul Israel's study of industrial research, *From Machine Shop to Industrial Laboratory: Telegraphy and the Changing Context of American Invention, 1830–1920* (Baltimore: Johns Hopkins University Press, 1992); and Andrew Dawson's study of the rise of systematized instruction, in "The Workshop and the Classroom: Philadelphia Engineering, the Decline of Apprenticeship, and the Rise of Industrial Training, 1878–1900," *History of Education Quarterly* 39 (1999): 143–60.

The Effects of Control

Gérard-Joseph Christian presided over the Conservatoire des Arts et Métiers during a crucial period in French industrialization. His theory of "Technonomie" suggested a way of assessing the performance of machines without quantifying it solely according to outputs and inputs; this chapter analyzes his theory of industrial technology as it bore on these concepts of machine performance and motion control. It considers Christian's numerous publications on industrial mechanics, and especially the fully developed statement of his theory in *Vues sur le système général des opérations industrielles, ou plan de technonomie* (Paris: Huzard and Courcier, 1819). Details of his career before he became director of the conservatory are available at the Archives Nationales, Paris (Report to Interior Minister, 1814, personnel dossier, Christian, G.-J., ser. F12, Conservatoire des Arts et Métiers, Conseil de Perfectionnement, Personnel, Matériel, Inventions, an IV–1879, file F12 4862). Two biographical sketches are held at the Centre d'Histoire des Techniques du Conservatoire National des Arts et Métiers, Paris (biographical notices for G.-J. Christian, 1880, 1903, doc. no. 2769). The manuscripts division of the CNAM's library holds several letters from Christian regarding the mounting of industrial exhibitions (Manuscript nos.

413–16, Library of the Conservatoire National des Arts et Métiers, Paris). Patent applications, including drawings, descriptions, and notes, are held in the archives of the Institut National de la Propriété Industrielle, Compiègne.

Essential is Antoine Picon's overview of Christian's career, "Gérard-Joseph Christian (1778–1832), directeur du Conservatoire (1816–1831)," in the very useful *Les Professeurs du Conservatoire National des Arts et Métiers: dictionnaire biographique, 1794–1955*, ed. Claudine Fontanon and André Grelon, 2 vols. (Paris: CNAM, 1994), 1:332–35. Jan Sebestik analyzed the concept of Technonomie in "De la technologie à la technonomie: Gérard-Joseph Christian," *Cahiers S.T.S.* 2 (1984): 56–69; Sebestik had earlier collaborated in an important study of the history of the concept of technology, in which Christian figures prominently, in Jacques Guillerme and Jan Sebestik, "Les Commencements de la technologie," *Thales* 12 (1966): 1–72. Christian's *Traité de mécanique industrielle* is available on microfilm, as is Leblanc's *Nouveau système complet de filature de coton*, both as part of the Goldsmiths'-Kress Library of Economic Literature, but their usefulness is seriously hampered, as neither microfilm includes the illustrating plates.

An outpouring of literature greeted the bicentennial of the CNAM. Claudine Fontanon and André Grelon's "Le Conservatoire des Arts et Métiers: deux siècles d'histoire" is an analytical overview (in *Les Professeurs du Conservatoire*, 23–57). Michel Le Moël and Raymond Saint-Paul edited a wide-ranging collection on the history of the institution, *Le Conservatoire National des Arts et Métiers au coeur de Paris* (Paris: CNAM, 1994): it includes Claudine Fontanon's "Conviction républicaine pour une fondation," on the political context of its founding and first decades; André Grelon's "Techniciens et ingénieurs," on its training functions through the interwar period of the twentieth century; Alain Mercier's "Le Portefeuille de Vaucanson," on the early composition of the drawing collection; and Dominique Ferriot's "Le Musée des Arts et Métiers," on the motivation for and pedagogical function of the CNAM's museum. Crafts and their relation to technical education were the subject of Walter Rathjen's "Wissen kann nicht mehr ein Privileg der Elite sein: 200 Jahre Conservatoire des Arts et Métiers in Paris" (*Kultur & Technik* 18 [1994]: 22–27). Robert Fox's study "Education for a New Age" (in *Artisan to Graduate: Essays to Commemorate the Foundation in 1824 of the Manchester Mechanics' Institution, Now in 1974 the University of Manchester Institute of Science and Technology*, ed. Donald S. L. Cardwell [Manchester: Manchester University Press, 1974], 24–38) analyzes the introduction of the lecture program as a contest between pedagogical traditionalists and liberals. Anatole Monzie's *Le Conservatoire de peuple* (Paris: Sedes, 1948) is an accessible, if dated, overview, and the National Ministry of Education published a useful volume with various timelines of who held which chairs, for example, and developments in different departments (*Cent-cinquante ans de Haut Enseignement Technique au Conservatoire National des Arts et Métiers* [Paris: National Ministry of Education, 1970]). *La Revue, Musée des Arts et Métiers*, a publication for general readers, often includes material of interest to historians—for example, Bruno Jacomy, "Les Dents et le mouvement," on the relationship between technical drawing, gearing, and motion (*La Revue* 15 [June 1996]: 37–44). Scholarship on various aspects of the institution's history appears in *Les Cahiers d'Histoire du CNAM*, including Louis André's biographical study of Leblanc, "César Nicolas Leblanc et le dessin de machines," *Les Cahiers d'Histoire du CNAM* 4 (1992): 71–92. A helpful look at sources for the history of the Conservatoire is Dominique Larroque's "Lieux et domaines de la recherche: le Centre d'Histoire des Techniques (CHDT) du Conservatoire National des Arts et Métiers," *Bulletin de l'Institute d'Histoire du Temps Présent* 45 (1991): 21–26. Leblanc's engravings are analyzed as arguments supporting state attempts to stimulate industrialization in Jennifer Karns Alexander, "The Line between Potential and Working Machines: César Nicolas Leblanc and Patent Engravings, 1811–1835," *History and Technology* 15 (1999): 175–212.

The history of technical education in France has similarly stimulated a large literature, especially concerning engineering education. In the last two decades historians have carefully distinguished between different traditions of engineering in France: an elite tradition centering on engineers trained at the École Polytechnique; a developing elite tradition for students trained at the École Centrale des Arts et Métiers; and an intermediate tradition, with students trained at the Écoles des Arts et Métiers. An excellent study treating technical education outside the École Polytechnique is James M. Edmonson, *From Mécanicien to Ingénieur: Technical Education and the Machine Building Industry in Nineteenth-Century France* (New York: Garland, 1987). C. R. Day treats technical education as a route of intergenerational social mobility in *The École d'Arts et Métiers and the Rise of French Industrial Education* (Cambridge, MA: MIT Press, 1987). John Hubbel Weiss treats the relationship between the École Centrale and the École Polytechnique in *The Making of Technological Man: The Social Origins of French Engineering Education* (Cambridge, MA: MIT Press, 1982); Weiss also addresses the changing professional structure of engineering in "Les Changements de structure dans la profession d'ingénieur en France de 1800 à 1850," in *L'Ingénieur dans la société française*, ed. André Thépot (Paris: Les Éditions Ouvrières, 1985). John Pannabecker analyzes a push for systematic instruction at the Écoles des Arts et Métiers in "School for Industry: L'École d'Arts et Métiers of Châlons-sur-Marne under Napoléon and the Restoration," *Technology and Culture* 43 (2002): 254–94. Bruno Belhoste analyzes the École Polytechnique as a national myth and an important structure in a developing technocratic system; see *La Formation d'une technocratie: l'École Polytechnique et ses élèves de la Révolution au Second Empire* (Paris: Belin, 2003). Engineers from the Polytechnique largely entered state service after having received additional training at one of the *grandes écoles;* Antoine Picon has analyzed the role of the prestigious school of roads and bridges in *L'Invention de l'ingénieur moderne: l'École des Ponts et Chaussées 1747–1851* (Paris: L'École Nationale des Ponts et Chaussées, 1992). Ivor Grattan-Guinness suggests that the heavily theoretical and mathematical approach of the Polytechnique was nonetheless useful to mechanics; his examples come from within the tradition of civil engineering, however, not from machine building. See Grattan-Guinness, "Work for the Workers: Advances in Engineering Mechanics and Instruction in France, 1800–1830," *Annals of Science* 41 (1984): 1–33.

Military engineering in France up to the Restoration has received particular attention in recent years. Ken Alder, in *Engineering the Revolution: Arms and Enlightenment in France, 1763–1815* (Princeton: Princeton University Press, 1997), demonstrates how thoroughly political were engineering and the production of engineering artifacts. Janis Langins, in *Conserving the Enlightenment: French Military Engineering from Vauban to the Revolution* (Cambridge, MA: MIT Press, 2004), describes military engineers as engaged in a conservative practice, though not reactionary, concerned with constructing a civilized society.

Technical education outside of engineering is addressed by André Grelon, who argues that the Restoration was a particularly important developmental period, in "La Mise en place des enseignements techniques sous las restauration," *Science et Techniques en Perspective*, ser. 2, 4 (2000): 221–33; and Goran Ahlstrom has usefully compared the relationship of industrial development to technical education in "Higher Technical Education and the Engineering Profession in France and Germany during the 19th Century," *Economy and History* 21 (1978): 51–88. An interesting and useful collection of official documents is the two-volume *L'Enseignement Technique de la Révolution à nos jours: textes officiels avec introduction, notes, et annexes* (Paris: Economica and Service d'Histoire de l'Éducation, 1987), ed. Thérèse Charmasson, Anne-Marie Lelorrain, and Yannick Ripa. Many French institutions of technical education were more closely allied with the state than was the case in other countries, a situation Andy Green addresses in "Technical Education and State Formation in 19th-century England and France," *History of Education* 24 (1994): 123–39. Frederick Artz's study remains the most thorough overview of institutions and practices; see *The Development of Technical Education in France, 1500–1850* (Cambridge, MA: Society for the History of Technology and MIT Press, 1966).

Much has been written on artisans and laborers in Restoration France, often with an eye toward identifying the source of the French labor movement. Gérard Noriel, in *Les Ouvriers dans la société française XIXe–XXe siècle* (Paris: Editions du Seuil, 1986), is particularly evocative of workers' worlds; and Abel Poitrineau, in *Ils travaillaient la France: métiers et mentalités du XVIe au XIXe siècle* (Paris: Armand Colin, 1992), provides a sociological analysis of work and behavior. Especially useful are Roger Magraw, *A History of the French Working Class* (Oxford: Blackwell, 1992); and William Hamilton Sewell, *Work and Revolution in France: The Language of Labor from the Old Regime to 1848* (Cambridge: Cambridge University Press, 1980). Olivier Marchand focuses on the labor market in *Le Travail en France, 1800–2000* (Paris: Nathan, 2000); and I. J. Prothero analyzes the connection between artisanal work and labor radicalism in *Radical Artisans in England and France, 1830–1870* (Cambridge: Cambridge University Press, 1997). Louis Chevalier's study of the overlapping images of workers and criminals, *Laboring Classes and Dangerous Classes, in Paris during the First Half of the Nineteenth Century* (New York: Howard Fertig, 1973), remains provocative.

On industrialization in France, see François Crouzet, *De la supériorité de l'Angleterre sur la France: l'économique et l'imaginaire* (Cambridge: Cambridge University Press, 1990); and the collection edited by Patrick Fridenson and André Straus, *Le capitalisme français XIX–XXe siècles* (Paris: Bayard, 1987). In two important early articles David Landes argued that the French lacked an entrepreneurial spirit: "French Entrepreneurship and Industrial Growth in the Nineteenth Century," *Journal of Economic History* 9 (1949): 45–61; and "French Business and the Business Man: A Social and Cultural Analysis," in *Modern France*, ed. E. M. Earle (Princeton: Princeton University Press, 1955). In *France and the Economic Development of Europe* (Princeton: Princeton University Press, 1961) Rondo Cameron rejected the idea that development in France had been retarded; and in *An Economic History of Modern France* (New York: Columbia University Press, 1979) François Caron argued that the French were indeed entrepreneurial. On political developments Guillaume de Bertier de Sauvigny's study *The Bourbon Restoration* (Philadelphia: University of Pennsylvania Press, 1966) remains a classic. René Rémond examines right-wing political thought in *Les droites en France* (Paris: Aubier Montaigne, 1982). A study by André Jardin and André-Jean Tudesq, *Restoration and Reaction, 1815–1848* (Cambridge: Cambridge University Press, 1983), provides a useful and careful chronology of political events and alignments. David Skuy has written an elegant and charming study of the reaction to the assassination of the duc de Berry, focusing on the miracle son born to his wife eight months later; see *Assassination, Politics, and Miracles: France and the Royalist Reaction of 1820* (Montreal: McGill-Queen's University Press, 2003).

A few classic, older texts are not only useful but also a pleasure to read. Ferdinand-Dreyfus's life of Rochefoucauld, *Un Philanthrope d'autrefois: La Rochefoucauld-Liancourt, 1747–1827* (Paris: Librairie Plon, 1903), is full of interesting details; Rochefoucauld served on the governing board of the CNAM until he was dismissed during the ministry of the Ultra-Royalists. Chaptal published a heavily statistical survey of French industry, *De l'industrie française* (Paris: Imprimerie Nationale Editions, 1993), in 1819, the same year Christian's *Plan de Technonomie* appeared. Charles Ballot's study of the introduction of machinery into industrializing France was published posthumously, completed by a committee from his notes and manuscripts; it contains surprising and lengthy quotations as well as attempts to update some of Chaptal's figures (*L'Introduction du machinisme dans l'industrie française* [Lille: O. Marquant and F. Rieder & Co., 1923]).

Economy of Nature

This chapter surveys uses of the concept of efficiency in two important publications: the famous first edition of Charles Darwin's *Origin of Species* (1859) and the eighth and final edition of Alfred Marshall's seminal *Principles of Economics: An Introductory Volume* (1920). A useful variorum edition traces the development of the *Principles* and is annotated by G. W. Guillebaud; see Alfred Marshall, *Principles of Economics*, 9th ed. (variorum) (London: Macmillan, for the Royal Economic Society, 1961).

Darwin exerts a perennial fascination on historians and the public alike. Two biographies stand out: Janet Browne's acclaimed two-volume treatment, *Charles Darwin: Voyaging* (Princeton: Princeton University Press, 1995), and *Charles Darwin: The Power of Place* (Princeton: Princeton University Press, 2002); and Adrian Desmond and James Moore's *Darwin: The Life of a Tormented Evolutionist* (New York: Norton, 1991). A facsimile of the first edition of the *Origin of Species* is readily available (Cambridge: Harvard University Press, 1964).

Darwin's principle of divergence has received renewed attention in recent debates between evolutionary theory and creationism; see Michael Ruse, *The Evolution-Creation Struggle* (Cambridge: Harvard University Press, 2005). Two earlier contributions come from Janet Browne, "Darwin's Botanical Arithmetic and the 'Principle of Divergence,'" *Journal of the History of Biology* 13 (1980): 53–89; and Silvan S. Schweber, "Darwin and the Political Economists: Divergence of Character," *Journal of the History of Biology* 13 (1980): 195–289. Robert Young addressed Darwin's language of agency in "Darwin's Metaphor: Does Nature Select?" *Darwin's Metaphor: Nature's Place in Victorian Culture* (Cambridge: Cambridge University Press, 1985).

The concept of the economy of nature has interested historians for some time. Donald Worster's study *Nature's Economy: A History of Ecological Ideas* (Cambridge: Cambridge University Press, 1994) remains a good introduction, although the focus is not on the history of biology. Philip Rehbock does look at biological conceptions in "John Fleming (1785–1857) and the Economy of Nature," in *From Linnaeus to Darwin: Commentaries on the History of Biology and Geology*, ed. Alwyne Wheeler and James H. Price, Papers from the Fifth Easter Meeting of the Society for the History of Natural History, "Natural History in the Early Nineteenth Century," March 28–31, 1983 (London: Society for the History of Natural History, 1985), 129–40; as does Frank N. Edgerton, in "Changing Concepts of the Balance of Nature," *Quarterly Review of Biology* 48 (1973): 322–50. Raymond L. Lindeman took up the issue of efficiency in biology in a classic study in the history of ecology, "The Trophic-Dynamic Aspect of Ecology," *Ecology* 23 (October 1942): 399–417.

Peter Groenewegen is the leading authority on Alfred Marshall, and his biography is especially strong on Marshall's political views and his role in reform at Cambridge University; see Groenewegen, *A Soaring Eagle: Alfred Marshall, 1842–1924* (Aldershot: Edward Elgar, 1995). John Whitaker has edited Marshall's correspondence in three volumes, *The Correspondence of Alfred Marshall, Economist* (Cambridge: Cambridge University Press, 1996); and a volume of Marshall's early papers, *Early Economic Writings of Alfred Marshall, 1867–1890* (London: Macmillan, 1975). Two volumes stand out among the many treatments of Marshall's economics for their detailed analysis of his theory in the context of politics and hopes for progress, both by David Reisman, *The Economics of Alfred Marshall* (London: Macmillan, 1986) and *Alfred Marshall: Progress and Politics* (New York: St. Martin's Press, 1987).

The classic study on biological thinking in economics is Jack Hirschleifer, "Economics from a Biological Viewpoint," *Journal of Law and Economics* 20 (1977): 1–52. Philip Mirowski's edited volume on economics and conceptions of nature is indispensable: *Natural Images in Economic Thought: Markets Read in Tooth & Claw* (Cambridge: Cambridge University Press, 1994). Margaret Schabas's contribution addresses the influence of Darwin on John Stuart Mill's and Marshall's thought, and has an excellent bibliography; see Schabas, "The Greyhound and the Mastiff: Darwinian Themes in Mill and Marshall," in Mirowski, *Natural Images in Economic Thought*, 322–35. Schabas has consistently placed Marshall's thought in the larger context of economic theory and the quest for scientific legitimacy; see *A World Ruled by Number: William Stanley Jevons and the Rise of Mathematical Economics* (Princeton: Princeton University Press, 1990) and *The Natural Origins of Economics* (Chicago: University of Chicago Press, 2005). Camille Limoges and Claude Ménard analyze biological metaphors in Marshall's work in terms of organization; see Limoges and Ménard, "Organization and the Division of Labor: Biological Metaphors at Work in Alfred Marshall's *Principles of Economics*," in Mirowski, *Natural Images in Economic Thought*, 336–59. Marshall's own reflective piece, "Mechanical and Biological Analogies in Economics," was reprinted in Alfred C. Pigou, ed., *The Memorials of Alfred Marshall* (London: Macmillan, 1925). An early piece on Darwin is discussed in Tiziano Rafaelli, "The Early Philosophical Writings of Alfred Marshall," *Research in the History of Economic Thought and Methodology* 4 (1994): supp., 53–159. Useful treatments of Marshall's analysis of industrial organization are D. P. O'Brien, "Marshall's Industrial Analysis," *Scottish Journal of Politi-*

cal Economy 37–38 (1990–91): 61–84; Lawrence Moss, "Biological Theory and Technological Entrepreneurship in Marshall's Writings," *Eastern Economic Journal* 7–8 (1981–82): 3–13; and Earl F. Beach, "Marshall's Theory of Industrial Progress," *Eastern Economic Journal* 7–8 (1981–82): 59–65. In *The Machinery Question and the Making of Political Economy, 1815–1848*, Maxine Berg examines the influence of technological developments on economics (Cambridge: Cambridge University Press, 1981); Gregory Claeys, *Machines, Money, and the Millennium: From Moral Economy to Socialism, 1815–1860* (Princeton: Princeton University Press, 1987), offers an analysis of socialist economic theory as influenced by technology. Philip Mirowski's widely cited work *More Heat Than Light: Economics as Social Physics: Physics as Nature's Economics* (Cambridge: Cambridge University Press, 1989) focuses on economics and physics.

Ben Marsden's study of Rankine and efficiency, "Engineering Science in Glasgow: Economy, Efficiency, and Measurement as Prime Movers in the Differentiation of an Academic Discipline," *British Journal for the History of Science* 25 (1992): 319–46, is seminal in my treatment, as is the important and ambitious series by Norton Wise and Crosbie Smith, "Work and Waste: Political Economy and Natural Philosophy in Nineteenth Century Britain," *History of Science* 27 (1989): 263–301; 27 (1989): 391–449; and 28 (1990): 222–61. A classic treatment of the discovery of the conservation of energy is Thomas S. Kuhn, "Energy Conservation as an Example of Simultaneous Discovery," in *Critical Problems in the History of Science,* ed. Marshall Clagett (Madison: University of Wisconsin Press, 1959), 321–56.

A substantial literature treats the development of thermodynamics and the laws of the conservation of energy. Donald S. L. Cardwell's study *From Watt to Clausius: The Rise of Thermodynamics in the Early Industrial Age* (Ithaca, NY: Cornell University Press, 1971) remains one of the best. Two additional articles by Cardwell are especially important to the study of efficiency: "Steam Engine Theory in the 19th Century: From Duty to Thermal Efficiency: From Parkes to Sankey," *Transactions of the Newcomen Society* 65 (1995–96), and "Some Factors in the Early Development of the Concepts of Power, Work, and Energy," *British Journal for the History of Science* 3 (1967): 209–44. Richard L. Hills has written about the notion of perfection in the work of James Watt; see Hills, "The Origins of James Watt's Perfect Engine," *Transactions of the Newcomen Society* 68 (1996–97). For a biographical approach with a good discussion of some of the main players, including Sadi Carnot, Joseph Fourier, James Joule, and William Thomson (Lord Kelvin), see David Lindley, *Degrees Kelvin: A Tale of Genius, Invention, and Tragedy* (Washington, DC: Joseph Henry Press, 2004). Crosbie Smith and M. Norton Wise's biography of Kelvin is dense but worth mining; see *Energy and Empire: A Biographical Study of Lord Kelvin* (Cambridge: Cambridge University Press, 1989). Accessible is Crosbie Smith, *The Science of Energy: A Cultural History of Energy Physics in Victorian Britain* (Chicago: University of Chicago Press, 1998). Robert Fox translated Sadi Carnot's study and provided an excellent introduction; see Carnot, *Reflexions on the Motive Power of Fire: A Critical Edition with the Surviving Scientific Manuscripts,* ed. and trans. Robert Fox (Manchester: Manchester University Press, 1976).

The idea of progress is of perennial interest to historians; the classic studies are J. B. Bury, *The Idea of Progress: An Inquiry into Its Origin and Growth* (New York and London: Macmillan, 1932); and Joseph Needham, *Integrative Levels: A Reevaluation of the Idea of Progress* (Oxford: Clarendon, 1937). The historiography is summarized up through 1980 in Georg Iggers, "Progress in Historiography and Social Thought," in *Progress and Its Discontents,* ed. Gabriel Almond, Marvin Chodorow, and Roy Harvey Pearce (Berkeley: University of California Press, 1982). Conceptions of progress in Victorian England are treated in Richard Drayton, *Nature's Government: Science, Imperial Britain, and the "Improvement" of the World* (New Haven: Yale University Press, 2000); and, with specific reference to industrial hopes, in Richard Bellon, "Science at the Crystal Focus of the World," in *Science in the Marketplace: Nineteenth-Century Sites and Experiences,* ed. Bernard Lightman and Aileen Fyfe (Chicago: University of Chicago Press, 2007).

Balance and Transformation

This chapter compares the idea of efficiency in the Progressive era as it was used in the leading engineering journal, *Engineering Magazine,* and in the prominent general interest journal most committed to personal efficiency, the *Independent.* Samuel P. Hays and Samuel Haber put efficiency on the historian's map in two classic studies of the Progressive era—Hays's of the conservation movement and the "gospel of efficiency," Haber's of Frederick Winslow Taylor, scientific management, and popular efficiency. Hays examined the technocratic roots of the conservation movement and emphasized a claim for managerial authority based on a platform of technical expertise; see Samuel P. Hays, *Conservation and the Gospel of Efficiency: The Progressive Conservation Movement, 1890–1920* (Cambridge: Harvard University Press, 1959). Haber corralled the whole vast landscape of Progressive era efficiency in an elegant, concise study, focusing on Taylor and his disciples and developing a taxonomy of four types of efficiency: mechanical, economic, social, and personal; see Samuel Haber, *Efficiency and Uplift: Scientific Management in the Progressive Era* (Chicago: University of Chicago Press, 1964). I have followed Haber's lead in looking to journals for evidence of how the concept of efficiency was constructed and used; Haber himself cited the *Independent* as being particularly devoted to personal efficiency. An important source on periodicals is Frank Luther Mott, *A History of American Magazines,* vol. 2 (Cambridge: Harvard University Press, 1938).

Neither Hays nor Haber focused on mechanical or technical efficiency although it played an important supporting role in both studies. Historical contributions that do address mechanical efficiency are not about the Progressive era; see Ben Marsden, "Engineering Science in Glasgow: Economy, Efficiency, and Measurement as Prime Movers in the Differentiation of an Academic Discipline," *British Journal for the History of Science* 25 (1992): 319–46; and D.S.L. Cardwell, "Steam Engine Theory in the 19th Century: From Duty to Thermal Efficiency, from Parkes to Sankey," *Transactions of the Newcomen Society* 65 (1993–94): 117–28. Haber's use of the term *mechanical efficiency* underscored his orientation outward and away from the workings of engines and machines, but it inadequately characterized a complex issue: mechanical efficiency had a very specific meaning at the time of which Haber wrote and was not simply the "energy output-input ratio of a machine." Helpful in this regard is J. G. Horner, *Dictionary of Terms Used in the Theory and Practice of Mechanical Engineering* (1936; rpt., London: Technical Press, 1960), which includes a large section devoted to historical meanings. P. M. Heldt's classic texts are *The*

Gasoline Motor (New York: Horseless Age, 1911) and *High-Speed Combustion Engines: Design: Production: Tests, 11th ed of The Gasoline Motor* (Nyack, NY: P. M. Heldt, 1941).

Taylor's system of scientific management lay outside the mechanical tradition; although he had been president of the American Society of Mechanical Engineers (ASME) and had gained renown for a series of metal-cutting innovations, the society refused to publish Taylor's seminal management papers. Bruce Sinclair's study of the ASME is insightful on these issues; see Sinclair, *The American Society of Mechanical Engineers: A Centennial History of the American Society of Mechanical Engineers, 1880–1980* (Toronto: University of Toronto Press, 1980). Monte Calvert's study of the transition from a shop to a school-trained culture in mechanical engineering is helpful in tracing Taylor's professional identification; see *The Mechanical Engineer in America, 1830–1910: Professional Cultures in Conflict* (Baltimore: Johns Hopkins Press, 1967). Brian Palmer argued long ago that an emphasis on Taylorism was misplaced and that the method had much less impact in use than as theory; see "Class, Conception and Conflict: The Thrust for Efficiency, Managerial Views of Labor and the Working Class Rebellion, 1903–22," *Review of Radical Political Economics* 7 (1975): 31–49. Essential works on Taylor are Daniel Nelson, *Frederick W. Taylor and the Rise of Scientific Management* (Madison: University of Wisconsin Press, 1980); and Robert Kanigel's biography, *The One Best Way: Frederick Winslow Taylor and the Enigma of Efficiency* (New York: Viking, 1997), which contains an excellent bibliography. Classic early studies of scientific management include Horace Bookwalter Drury, *Scientific Management: A History and Criticism* (New York: Columbia University Press, 1922); and L. Urwick and E. F. L. Brech's multivolume study, *The Making of Scientific Management* (London: Sir Isaac Pitman and Sons, 1945–57); the third volume is a study of the famous experiments in human productivity at the Hawthorne plant of the Western Electric Company of Chicago from 1927 to 1932.

Hays wrote mostly about civil engineers, for whom efficiency had a less technical and more general meaning, more closely tied to political or social issues, such as water engineering and public health, than to the engines and machines at efficiency's mechanical core. Martin Melosi, Clayton McShane, and S. K. Schultz similarly treat efficiency in the context of sanitary engineers and city design more closely allied with civil engineering and public works than with mechanical engineering and heat engines; see Martin V. Melosi, *Garbage in the Cities: Refuse, Reform, and the Environment, 1880–1980* (College Station: Texas A&M University Press, 1981), esp. chap. 3; and S. K. Schultz and C. McShane, "To Engineer the Metropolis: Sewers, Sanitation, and City Planning in Late-Nineteenth-Century America," *Journal of American History* 65 (1978): 389–411. Amy Slaton's study of expertise and forms of specialized knowledge in building concentrates on the Progressive era; see *Reinforced Concrete and the Modernization of American Building, 1900–1930* (Baltimore: Johns Hopkins University Press, 2001). Edwin T. Layton's study of the American Society of Civil Engineers, *The Revolt of the Engineers: Social Responsibility and the American Engineering Profession* (1971; rpt., Baltimore: Johns Hopkins University Press, 1986), is a model of political and social analysis of technology.

An inadequate understanding of technical efficiency has led some authors to treat other than technical uses of efficiency as slogans, metaphors, or theories, rather than as serious attempts to apply a powerful and generalizable concept to an extended variety of human situations; see Cecelia Tichi, *Shifting Gears: Technology, Literature, Culture in Modernist America* (Chapel Hill: University of North Carolina Press, 1987), 63; Martha Banta, *Taylored Lives: Narrative Productions in the Age of Taylor, Veblen, and Ford* (Chicago: University of Chicago Press, 1993), ix; and John M. Jordan, *Machine-Age Ideology: Social Engineering and American Liberalism, 1911–1939* (Chapel Hill: University of North Carolina Press, 1994), 37, 44. Haber himself used the term *craze* to describe widespread attention to efficiency; *craze* suggests mania or rage, common enough synonyms for a fad, a fashion, or a style, but the underlying note remains one of disturbance or unbalance. In "The Business of Selling Efficiency: Harrington Emerson and the Emerson Efficiency Engineers, 1900–1930" (Ph.D. diss., Pennsylvania State University, 1992), James Parvin Quigel Jr. suggests that treating popular or personal efficiency as a craze has distorted our understanding of the reach and serious purpose of Progressive era interests in efficiency.

L. P. Alford's early biography of Henry Gantt is a good read, *Henry Laurence Gantt: Leader in Industry* (New York: American Society of Mechanical Engineers, 1934). Gantt published a number of works himself, including *Work, Wages, and Profits: Their Influence on the Cost of Living* (New York: Engineering Magazine, 1910) and *Industrial Leadership, Page Lectures, 1915, before the Sheffield Scientific School* (New Haven: Yale University Press, 1916). An excellent study of Harrington Emerson is Quigel, "Business of Selling Efficiency"; see also "Harrington Emerson," *National Cyclopedia of American Biography* (New York: James T. White, 1916); and Emerson, *Efficiency as a Basis for Operation and Wages* (New York: Engineering Magazine, 1911). Edward Earle Purinton's books read like modern guides to self-management; see *Efficient Living* (New York: Robert H. McBride, 1915) and *Triumph of the Man Who Acts* (New York: Robert H. McBride, 1916).

Merritt Roe Smith called on historians of technology to integrate their work with labor history, and the Progressive era provides a good context for such an effort ("Industry, Technology, and the 'Labor Question' in 19th Century America: Seeking Synthesis," *Technology and Culture* 32 [1991]: 555–70). Studies of workers include Daniel T. Rodgers, *The Work Ethic in Industrial America, 1850–1920* (Chicago: University of Chicago Press, 1978); and Daniel Nelson, *Managers and Workers: Origins of the New Factory System in the United States 1880–1920* (Madison: University of Wisconsin Press, 1975). David Montgomery has analyzed workers as historical agents in their own right and workers and the labor movement, from the nineteenth century into the twentieth; among his books addressing the Progressive era are *The Fall of the House of Labor: The Workplace, the State, and American Labor Activism, 1865–1925* (Cambridge: Cambridge University Press; Paris: Éditions de la Maison des Sciences de l'Homme, 1987) and *Workers' Control in America: Studies in the History of Work, Technology, and Labor Struggles* (Cambridge: Cambridge University Press, 1979). Montgomery also did a very provocative study of the interrelated historical conceptions of labor, wages, and political freedom, in *Citizen Worker: The Experience of Workers in the United States with Democracy and the Free Market during the Nineteenth Century* (Cambridge: Cambridge University Press, 1993). Another provocative and well-known study, but for a later time period, is David F. Noble, *America by Design: Science, Technology, and the Rise of Corporate Capitalism* (1977; rpt., Oxford: Oxford University Press, 1979).

Two classic studies analyzed the Progressive period as a whole and set the tone for much research that followed: Richard Hofstadter, in *The Age of Reform: From Bryan to FDR* (New

York: Vintage, 1955), made reform a central theme for historians; and Robert H. Wiebe, in *The Search for Order, 1877–1920* (New York: Hill and Wang, 1967), found not only enthusiasm for reform and change but an underlying trauma as well. Change is the central theme of John Whiteclay Chambers II's study *The Tyranny of Change: America in the Progressive Era, 1890–1920* (New Brunswick: Rutgers University Press, 1992). Daniel T. Rodgers discusses the problem of the very conception of progressivism, in "In Search of Progressivism," *Reviews in American History* 10 (1982): 113–32. A good guide to current thinking about the Progressive era may be found in Lewis L. Gould, *America in the Progressive Era, 1890–1914* (Essex: Pearson Education Limited, 2001); it deals primarily with political history, but the annotated bibliography is much broader and especially helpful (112–21). Another important study is David W. Noble, *The Progressive Mind, 1890–1917*, rev. ed. (Minneapolis: Burgess, 1981). Terry Smith analyzed the complex interplay of technology, aesthetics, and modernity in *Making the Modern: Industry, Art, and Design in America* (Chicago: University of Chicago Press, 1993). Rodgers, in a much-discussed recent book, portrayed progressivism as an international movement of reconstruction and reform; see *Atlantic Crossings: Social Politics in a Progressive Age* (Cambridge, MA: Belknap Press, 1998). A critical review was written by Thomas Haskell, who found the book important although he believed its thesis was flawed ("Taking Exception to Exceptionalism: Daniel Rodgers, *Atlantic Crossings*," *Reviews in American History* 28 [March 2000]: 151–66).

Historians have generally located the United States's and Britain's particular zeal for efficiency in the decades leading up to World War I; reform efforts centering on efficiency in France and Germany had significant presence after the war and through the 1920s. See G. R. Searle, *The Quest for National Efficiency: A Study in British Politics and Political Thought, 1899–1914* (1971; rpt., London: Ashfield Press, 1990); Sheila Faith Weiss, *Race Hygiene and National Efficiency: The Eugenics of Wilhelm Stallmayer* (Berkeley: University of California Press, 1987); and Mary Nolan, *Visions of Modernity: American Business and the Modernization of Germany* (New York: Oxford University Press, 1994).

An Island of Mechanical Predictability

This chapter investigates a traveling exhibit on worker seating in the late years of the Weimar Republic, as described in journals and reports on rationalization and labor physiology, especially the *Zentralblatt für Gewerbehygiene und Unfallverhütung*, the *RKW-Nachrichten* of the Reichskuratorium für Wirtschaftlichkeit (RKW), and the *Siemens Jahrbuch* of the Siemens concerns. Documentation of an exhibit on "The Human and Rationalization" in Düsseldorf in 1930, including photographs and a catalog, is held at the Bundesarchiv Berlin ("Die deutsche Rationalisierungsbewegung im Reichsmuseum-Düsseldorf," Bundesarchiv Berlin, R 107/1). Materials are more abundant for the RKW after 1933—for example, in "Reichskuratorium für Wirtschaftlichkeit, Organisation und Verwaltung," covering 1933–39 (file R 107/2, Bundesarchiv Berlin).

The late 1920s was a volatile time in Germany, and a voluminous literature on both the end of the Weimar Republic and the rationalization movement exists. Richard J. Evans includes an extensive bibliography in *The Coming of the Third Reich* (New York: Penguin, 2004), and although the book does not discuss rationalization except in passing, it is an essential source of information about complicated social, cultural, and political developments. Detlev J. Peukert discussed rationalization more extensively in the context of politics and society in the crisis years, in *The Weimar Republic: The Crisis of Classical Modernity* (New York: Hill and Wang, 1993). Hans Mommsen's *The Rise and Fall of Weimar Democracy* (Chapel Hill: University of North Carolina Press, 1996) analyzes the political fortunes and tragedies of the republican experiment. Jeffrey Herf named an important Weimar phenomenon in *Reactionary Modernism: Technology, Culture, and Politics in Weimar and the Third Reich* (Cambridge: Cambridge University Press, 1984).

Carola Sachse's study of Siemens, *Betriebliche Sozialpolitik als Familienpolitik in der Weimarer Republik und im Nationalsozialismus: Mit einer Fallstudie über die Firma Siemens, Berlin* (Hamburg: Hamburger Institut füt Sozialforschung, 1987), provides a dual context of worker interests and business management, as does J. Ronald Shearer, in two pieces: "Talking about Efficiency: Politics and the Industrial Rationalization Movement in Weimar Germany," *Central European History* 28 (1995): 483–506; and "The Politics of Industrial Efficiency in the Weimar Republic: Technological Innovation, Economic Efficiency, and Their Social Consequences in the Ruhr Coal Mining Industry, 1918–1929" (Ph.D. diss., University of Pennsylvania, 1989). Charles Maier's well-known consideration of rationalization between the wars remains important; see "Zwischen Taylorismus und Technokratie," in *Die Weimarer Republik, Belagerte Civitas*, ed. Michael Stürmer (Königstein: Athenäum, 1980), 188–213. The development of ideas about work in Germany is discussed by Joan Campbell in *Joy in Work, German Work: The National Debate, 1800–1945* (Princeton: Princeton University Press, 1989).

Important treatments of Weimar rationalization include Mary Nolan's study of Germany's importation of American ideas of productivity, especially Fordism, in *Visions of Modernity: American Business and the Modernization of Germany* (New York: Oxford University Press, 1994); Heidrun Homburg's work on management strategy in "The 'Human Factor' and the Limits of Rationalization: Personnel Management Strategies and the Rationalization Movement in German Industry between the Wars," in *The Power to Manage? Employers and Industrial Relations in Comparative Historical Perspective*, ed. Steven Tolliday and Jonathan Zeitlin (London: Routledge, 1991), 147–75; and Homburg, *Rationalisierung und Industriearbeit: Arbeitsmarkt-Management-Arbeiterschaft im Siemens-Konzern Berlin, 1900–1939* (Berlin: Haude & Spener, 1991). Wilfried Feldenkirchen's business history of Siemens is helpful (*Siemens, 1918–1945* [Columbus: Ohio State University Press, 1999]). Hans Wupper-Tewes examines the political and disciplinary aspects of rationalization in *Rationalisierung als Normalisierung: Betriebswissenschaft und betriebliche Leistungspolitik in der Weimarer Republik* (Münster: Westfälisches Dampfboot, 1995). The contemporary analysis of the American business critic Robert A. Brady remains relevant; see Brady, *The Rationalization Movement in German Industry: A Study in the Evolution of Economic Planning* (Berkeley: University of California Press, 1933) and "The Meaning of Rationalization: An Analysis of the Literature," *Quarterly Journal of Economics* (May 1932): 526.

The RKW exhibit was part of an international tradition of exhibits on worker safety and health; more than a decade earlier Frank B. Gilbreth and Lillian M. Gilbreth had opened a fatigue museum in Providence, Rhode Island, principally displaying advances in the design of workers' chairs; see Gilbreth and Gilbreth, *Fatigue Study* (New York: Sturgis and

Walton, 1916), 99–113; see also Stefan Poser, *Museum der Gefahren, Die gesellschaftliche Bedeutung der Sicherheitstechnik: Das Beispiel der Hygiene-Ausstellungen und Museum für Arbeitsschutz in Weim, Berlin und Dresden* (Münster, 1998).

A history of ergonomics and human factors accounting for its broad range of antecedents has yet to be written. Influences included anthropology, anthropometry, physiology, anatomy, and psychology as well as developments in physics and mechanics. Anson Rabinbach has described a number of these contributions in *The Human Motor: Energy, Fatigue, and the Origins of Modernity* (Berkeley: University of California Press, 1990). Two studies chart the terrain: Gertraud Schottdorf, *Arbeits-und Leistungsmedizin in der Weimarer Republik* (Husum: Matthiesen Verlag, 1995) and Martin Höfler-Waag, *Die Arbeits-und Leistungsmedizin im Nationalsozialismus von 1939–1945* (Husum: Matthiesen Verlag, 1994). Rainer-W. Hoffman analyzes the German science of work in *Wissenschaft und Arbeitskraft: Zur Geschichte der Arbeitsforschung in Deutschland* (Frankfurt: Campus Verlag, 1985), and a recent and fascinating study of dance and mechanical movement comes from Christiane Kuhlmann, *Bewegter Körper—Mechanischer Apparat* (Frankfurt am Main: Peter Lang, 2001). The influence of gymnastics, dance, sport, and physical therapy has been little explored, nor has the influence of gender and concepts of women's work. Histories written by ergonomic practitioners have concentrated on developments since World War II; see David Meister, *The History of Human Factors and Ergonomics* (Mahwah, NJ: Lawrence Erlbaum Associates, 1999), 146–230; Fariborz Tayyari and James L. Smith, *Occupational Ergonomics: Principles and Applications* (London: Chapman & Hall, 1997), 1–10; K. H. E. Kroemer, H. J. Kroemer, and K. E. Kroemer-Elbert, *Engineering Physiology: Bases of Human Factors/Ergonomics*, 3rd ed. (New York: Van Nostrand Reinhold, 1997), 1–53.

On engineering, a useful source is Kees Gispen, *New Profession, Old Order: Engineers and German Society, 1815–1914* (Cambridge: Cambridge University Press, 1989); Hans-Joachim Braun and Wolfhard Weber discuss Franz Reuleaux in "Ingenieurwissenschaft und Gesellschaftspolitik: Das Wirken von Franz Reuleaux," in *Wissenschaft und Gesellschaft: Beiträge zur Geschichte der Technischen Universität Berlin, 1879–1979*, ed. Reinhard Rürup (Berlin: Springer Verlag, 1979), 1:285–300.

There are many additional and useful studies of rationalization. In "Politische Krise und Rationalisierungsdiskurs in den zwanziger Jahren," *Technikgeschichte* 62 (1995): 317–22, Gunther Mai emphasizes crisis. Peter Hinrichs's longer-range study looks at the Weimar period as one of consolidation of gains made in industrial psychology; see *Um die Seele des Arbeiters: Arbeitspsychologie, Industrie-u. Betriebssoziologie in Deutschland, 1871–1945* (Cologne: Pahl-Rugenstein, 1981). See also Thomas von Freyberg, *Industrielle Rationalisierung in der Weimarer Republik* (Frankfurt: Campus Verlag, 1989); Gunnar Stollberg, *Die Rationalisierungsdebatte 1908–1933* (Frankfurt: Campus Verlag, 1981); and, more generally, Jürgen Kocka, *Industrial Culture and Bourgeois Society: Business, Labor, and Bureaucracy in Modern Germany* (New York: Berghahn Books, 1999).

Rationalization's gendered character has been of interest to historians in recent years; see Mary Nolan, "'Housework Made Easy': The Taylorized Housewife in Weimar Germany's Rationalized Economy," *Feminist Studies* 16 (1990): 549–78; Barbara Orland, "Effizienz im Heim: Die Rationalisierungsdebatte zur Reform der Hausarbeit in der Weimarer Republik," *Kultur und Technik* 4 (1983): 221–27; and Helgard Kramer, "'Rationelle Haushaltsführung' und die 'neue Frau' in der Zwanziger Jahre," *Feministische Studien* 1 (1982): 123–26.

Sheila Faith Weiss discusses efficiency and eugenics in *Race Hygiene and National Efficiency: The Eugenics of Wilhelm Schallmayer* (Berkeley: University of California Press, 1987); by *efficiency* she means the rational management of population. See also Paul Weindling, *Health, Race and German Politics between National Unification and Nazism, 1870–1945* (Cambridge: Cambridge University Press, 1989); and Erwin Baur, Eugen Fischer, and Fritz Lenz, *Human Heredity*, trans. Eden Paul and Cedar Paul (New York: Macmillan, 1931). On the uniqueness of German eugenics, see Paul Weindling, "The Sonderweg of German Eugenics: Nationalism and Scientific Internationalism," *British Journal for the History of Science* 22 (1989): 321–33; Peter Weingart, "German Eugenics between Science and Politics," *Osiris* 5 (1989): 260–82; and Jonathan Harwood, "National Styles in Science: Genetics in Germany and the United States between the World Wars," *Isis* 78 (1987): 390–414.

The literature on unemployment in Weimar is also vast. Labor and street violence are analyzed by Eve Rosenhaft in *Beating the Fascists? The German Communists and Political Violence, 1929–1933* (Cambridge: Cambridge University Press, 1983). Alex Zubas demonstrates that employers worried about unrest among the unemployed; see "Lazy, Apathetic, and Dangerous: The Social Construction of Unemployed Workers in Germany during the Late Weimar Republic," *Contemporary European History* 10 (2001): 25–49. Peter D. Stachura provides statistical information in "The Development of Unemployment in Modern German History," in *Unemployment and the Great Depression in Weimar Germany*, ed. Peter D. Stachura (New York: St. Martin's Press, 1986); see also Hans-Ulrich Thamer, *Verführung und Gewalt: Deutschland, 1933–1945* (Berlin: Siedler, 1986). Rationalization was heavily blamed for unemployment; how accurate that charge was is investigated in Eva Cornelia Schöck, *Arbeitslosigkeit und Rationalisierung* (Frankfurt: Campus Verlag, 1977). David F. Crew looks at the effects of unemployment on life situations and morale, in *Germans on Welfare: From Weimar to Hitler* (New York: Oxford University Press, 1998); Young-Sun Hong returns to the question of Weimar modernity in *Welfare, Modernity, and the Weimar State, 1919–1933* (Princeton: Princeton University Press, 1998).

Pride in Efficiency

The evolution of Robert William Fogel's thought on efficiency and slavery is the subject of this chapter, which considers his published work. The evolution of Fogel's thought can be seen in his two main volumes: *Time on the Cross: The Economics of American Negro Slavery* (1974; rpt., New York: Norton, 1995) and *Without Consent or Contract: The Rise and Fall of American Slavery* (New York: Norton, 1989). Much of the data and supporting analysis can be found in the companion volumes, although they have been faulted for being incomplete and difficult to use; see Fogel and Engerman, eds., *Time on the Cross: Evidence and Methods, A Supplement* (Boston: Little, Brown, 1974); and the three supporting volumes for *Without Consent or Contract*: Fogel, Ralph A. Galantine, and Richard L. Manning, eds., *Without Consent or Contract: The Rise and Fall of American Slavery: Evidence and Methods* (New York: Norton, 1992); Fogel and Engerman, eds., *Without Consent or Contract: The Rise and Fall of American Slavery: Markets and Production: Technical Papers*, vol. 1 (New York: Norton, 1992); and Fogel and Engerman, eds., *Without Consent or Contract: The Rise and Fall of American Slavery: Conditions of Slave Life and the Transition to Freedom: Technical Papers*, vol. 2 (New York: Norton, 1992). Many of the technical papers in the supplementary volumes have also been published elsewhere. Fogel's Fleming Lectures at Louisiana State University provide an accessible overview of how he now views the disputes; see Fogel, *The Slavery Debates: A Retrospective, 1952–1990* (Baton Rouge: Louisiana State University Press, 2003). For some years Fogel has been working on questions of demography and human nutrition in history, an outgrowth of the work on slavery; he recently pioneered the term *technophysio evolution* to describe the sources of advances in health and longevity in the widely quoted work *The Escape from Hunger and Premature Death, 1700–2100: Europe, America, and the Third World* (Cambridge: Cambridge University Press, 2004).

Thomas L. Haskell wrote one of the earliest and most accessible of the critiques of *Time on the Cross*, "Were Slaves More Efficient? Some Doubts about *Time on the Cross*," *New York Review of Books*, September 19, 1974, 38–42 (it also appeared in Haskell's collected essays, *Objectivity Is Not Neutrality: Explanatory Schemes in History* [Baltimore: Johns Hopkins University Press, 1998], 31–39). A number of technical papers critical of *Time on the Cross* were collected in Paul A. David, Richard Sutch, Peter Temin, and Gavin Wright, eds., *Reckoning with Slavery: Critical Essays in the Quantitative History of American Negro Slavery* (New York: Oxford University Press, 1976); Kenneth M. Stampp's "Introduction: A Humanistic Perspective" reviews more traditional historical evidence. Another critical volume is Herbert Gutman, *Slavery and the Numbers Game: A Critique of* Time on the Cross (Urbana: University of Illinois Press, 1975); Gutman's work is more accessible than much of the literature from economic historians, and he made it a point to assist the nonspecialist historian to interpret *Time on the Cross*. A special volume on the dispute was edited by Gar Walton, "A Symposium on *Time on the Cross*," *Explorations in Economic History* 12 (1975). Thomas Haskell's review of these volumes provides a good overview of the debate; see "The True and Tragical History of *Time on the Cross*," *New York Review of Books*, October 2, 1974, 33–39 (it also appears in Haskell, *Objectivity Is Not Neutrality*, 40–56). The *American Economic Review* published an exchange including Fogel and Engerman, "Explaining the Relative Efficiency of Slave Agriculture in the Antebellum South," *American Economic Review* 67 (1977): 275–96; Paul A. David and Peter Temin, "Explaining the Relative Efficiency of Slave Agriculture in the Antebellum South: Comment," *American Economic Review* 69 (1979): 213–18; Gavin Wright, "The Efficiency of Slavery: Another Interpretation," *American Economic Review* 69 (1979): 219–26; D. F. Schaefer and M. D. Schmitz, "The Relative Efficiency of Slave Agriculture: A Comment," *American Economic Review* 69 (1979): 208–12; Thomas Haskell, "Explaining the Relative Efficiency of Slave Agriculture in the Antebellum South: A Reply to Fogel and Engerman," *American Economic Review* 69 (1979): 206–7; and Fogel and Engerman, "Explaining the Relative Efficiency of Slave Agriculture in the Antebellum South: Reply," *American Economic Review* 70 (1980): 672–90. Hugh G. J. Aitken edited a collection of articles on the question of slavery's efficiency just a few years before *Time on the Cross* was published; see *Did Slavery Pay? Readings in the Economics of Black Slavery in the United States* (Boston: Houghton Mifflin, 1971).

Recent studies of the slave economy include Roger L. Ransom, *Conflict and Compromise: The Political Economy of Slavery, Emancipation, and the American Civil War* (Cambridge: Cambridge University Press, 1989); Michael Tadman, *Speculators and Slaves: Masters, Traders, and Slaves in the Old South* (Madison: University of Wisconsin Press, 1989); and James L. Huston, *Celebrating the Value of the Union: Slavery, Property Rights, and the Economic Origins of the Civil War* (Chapel Hill: University of North Carolina Press, 2003).

A number of studies of slavery as a labor system in the antebellum South have achieved the status of classics in the history of the antebellum United States, not because they remain unchallenged but because they have cast a long shadow on ensuing interpretations. Among them are Ulrich Bonnell Phillips, *American Negro Slavery: A Survey of the Supply, Employment and Control of Negro Labor as Determined by the Plantation Regime* (1918; rpt., Gloucester, MA: Peter Smith, 1959), whose interpretation of a benign management of slave labor was challenged by Kenneth M. Stampp in *The Peculiar Institution: Slavery in the Ante-Bellum South* (New York: Knopf, 1956), who drew a stark picture of slave resistance; and by Stanley M. Elkins, who likened living under slavery to being in a Nazi concentration camp, in *Slavery: A Problem in American Institutional and Intellectual Life* (Chicago: University of Chicago Press, 1959). Eugene D. Genovese's studies of southern society and the economy have also achieved classic status; see *The Political Economy of Slavery: Studies in the Economy and Society of the Slave South* (New York: Random House, 1967) and *Roll, Jordan, Roll: The World the Slaves Made* (New York: Pantheon Books, 1974).

The relationship between slavery and technology, and technological change, was addressed by R. Keith Aufhauser in "Slavery and Technological Change," *Journal of Economic History* 34 (1974): 36–50; and more recently by A. Glenn Crothers in "Agricultural Improvement and Technological Innovation in a Slave Society: The Case of Early National Northern Virginia," *Agricultural History* 75 (2001): 135–67; and Stanley Engerman, in "Contract Labor, Sugar, and Technology in the Nineteenth Century," *Journal of Economic History* 43 (1983): 635–59. The importance of American slavery to industrialization has been explored by Ronald Bailey, "The Other Side of Slavery: Black Labor, Cotton, and the Textile Industrialization of Great Britain and the United States," *Agricultural History* 68 (1994): 35–50; Sven Beckert, "Emancipation and Empire: Reconstructing the Worldwide Web of Cotton Production in the Age of the American Civil War," *American Historical Review* 109 (2004): 1405–38; Michael V. Kennedy, "The Hidden Economy of Slavery: Commercial and Industrial Hiring in Pennsylvania, New Jersey, and Delaware, 1728–1800," *Essays in Economic and Business History* 21 (2003): 115–25; and William Darity Jr., "The Numbers Game and the Profitability of the British Trade in Slaves," *Journal of Economic History* 45 (1985): 693–703. David Eltis argued that for the sugar industry the slave trade did not appear to have been critical in terms of industrialization, in "The Importance of Slavery and the Slave Trade to Industrializing Britain," *Journal of Economic History* 60 (2000): 123–44.

Slavery in international perspective was addressed by Orlando Patterson in a magisterial study, *Slavery and Social Death: A Comparative Study* (Cambridge: Harvard University Press, 1982); and the distinguished scholar David Brion Davis has written a number of essential studies, most recently *In Human Bondage: The Rise and Fall of Slavery in the New World* (Oxford: Oxford University Press, 2005); see also *In the Image of God: Religion, Moral Values, and Our Heritage of Slavery* (New Haven: Yale University Press, 2001); *The Problem of Slavery in the Age of Revolution, 1770–1823* (Oxford: Oxford University Press, 1999); *Slavery and Human Progress* (Oxford: Oxford University Press, 1984); and *The Problem of Slav-*

ery in Western Culture (Ithaca, NY: Cornell University Press, 1966). Davis, Thomas L. Haskell, and John Ashworth engaged in a debate on the relationship between capitalism and slavery in the *American Historical Review;* the papers are collected, and two previously unpublished pieces added, in Ashworth, Davis, and Haskell, *The Antislavery Debate: Capitalism and Abolitionism as a Problem in Historical Interpretation* (Berkeley: University of California Press, 1992).

Global Efficiency

This chapter surveys three unlikely places to look for efficiency in contemporary society, against the backdrop of the increasing pressures of a globalized workplace: in lawsuits over free speech, business interest in early childhood education, and attempts to manage disruptions through computing. The examples were chosen for the serious connections they make between efficiency and high-order values such as social equity, freedom of expression, and freedom to work.

The literature on globalization is vast. An interesting contrast may be made between Thomas L. Friedman's cheerleading in *The World Is Flat: A Brief History of the Twenty-first Century* (New York: Farrar, Straus and Giroux, 2005) and Joseph E. Stiglitz's measured and critical assessment in *Globalization and Its Discontents* (2002; rpt., New York: Norton, 2003). Particular attention is paid to the global trading system in the wake of the protests against the World Trade Organization in Seattle in the collection of essays *Efficiency, Equity, Legitimacy: The Multilateral Trading System at the Millennium,* ed. Roger B. Porter, Pierre Sauvé, Arvind Subramanian, and Americo Beviglia Zampetti (Washington, DC: Brookings Institution Press, 2001). Issues of the global economy should be studied in the context of local and national economies; a good place to begin is with the work of Nick Tiratsoo and Jim Tomlinson on productive efficiency and relative economic decline in Britain. See Tiratsoo and Tomlinson, *Industrial Efficiency and State Intervention: Labour 1939–51* (London: Routledge, 1993) and *The Conservatives and Industrial Efficiency, 1951–64: Thirteen Wasted Years?* (London: Routledge, 1998).

Daniel Bell's study *The Coming of Post-Industrial Society: A Venture in Social Forecasting* (New York: Basic Books, 1973) has become the classic text of the phenomenon of globalization. Steven Best and Douglas Kellner explore the effect on personal identity of the new technocapitalism, arguing that it requires people to master new technologies of communication if they wish to participate in the global public sphere; see Best and Kellner, *The Postmodern Adventure: Science, Technology, and Cultural Studies at the Third Millennium* (New York: Guilford Press, 2001), esp. chap. 5: "Globalization and the Restructuring of Capital," 205–53. Margaret A. Rose provides a useful analysis of the relationship between theories of the postmodern and of the postindustrial in *The Post-Modern and the Post-Industrial: A Critical Analysis* (Cambridge: Cambridge University Press, 1991). An essential critical analysis, centering on theories of the information society, post-Fordism, and postmodernity, is the second edition of Krishan Kumar, *From Post-Industrial to Post-Modern Society: New Theories of the Contemporary World* (Malden, MA: Blackwell, 2005). A good deal about the values attached to conceptions of the postindustrial may be gleaned from studies intended to advise managers on how to guide their businesses or other organizations in responding; for examples see Larry Hirschhorn, *Beyond Mechanization: Work and Technology in a Postindustrial Age* (Cambridge, MA: MIT Press, 1984); and Charles Heckschier, Michael Maccoby, Rafael Ramírea, and Pierre-Eric Tixier, *Agents of Change: Crossing the Post-Industrial Divide* (Oxford: Oxford University Press, 2003).

On efficiency and the law an important distinction must be made between efficiencies in the legal process and legal precedents regarding efficiency in the workplace or other sites of productive activity. On the legal process Thane Rosenbaum offers compelling reflections in *The Myth of Moral Justice: Why Our Legal System Fails to Do What's Right* (New York: HarperCollins, 2004); a good excerpt appears in *Tikkun* 19 (May–June 2004): 65–70. I was led to *Connick* and *Pickering*, the cases treating efficiency in governmental workplaces, by work on academic freedom and free speech. Jeffrey S. Strauss analyzes the legal relationship between public concern for efficiently executed governmental functions and disruptive free speech in "Dangerous Thoughts? Academic Freedom, Free Speech, and Censorship Revisited in a Post–September 11th America," *Washington University Journal of Law and Policy* 15 (2004): 343–67. The tension between political and economic freedom and the instrumental logic of management are the subjects of Rebecca Gose Lynch, in "Pawns of the State or Priests of Democracy? Analyzing Professors' Academic Freedom Rights within the State's Managerial Realm," *California Law Review* 91 (2003): 1061–1108. Risa L. Lieberwitz analyzes the contrast between corporate and academic models of teaching and their management and free speech implications in "The Corporatization of the University: Distance Learning at the Cost of Academic Freedom?" *Boston University Public Interest Law Journal* 12 (2002): 73–135. Mention should also be made of the emphasis on efficiency in the law, particularly how efficient distribution of resources has been achieved by the common law, in the work of members of the Chicago school of law and economics. A good example may be found in Richard A. Posner, *Economic Analysis of Law,* 3rd ed. (Toronto: Little, Brown, 1983).

Educational policy raises the issue of efficiency versus equity, a dichotomy long recognized in economic thought. Essential to understanding the construction of this duality is Arthur M. Okin's concise and straightforward essay, *Equality and Efficiency: The Big Tradeoff* (Washington, DC: Brookings Institution, 1975). The interest in efficiency in education is not new. Raymond E. Callahan's classic historical study locates its roots early in the twentieth century; see *Education and the Cult of Efficiency: A Study of the Social Forces That Have Shaped the Administration of the Public Schools* (Chicago: University of Chicago Press, 1962). The legacy of Callahan's work, in historical scholarship as well as in school management and organization, is discussed in a collection edited by William Edward Eaton, *Shaping the Superintendency: A Reexamination of Callahan and* The Cult of Efficiency (New York: Teachers College Press, 1990). A broader context for the issue, placed both in a larger geographic framework and a longer time frame, appears in the essays collected by editors Robert Fox and Anna Guagnini, in *Education, Technology, and Industrial Performance in Europe, 1850–1939* (Cambridge: Cambridge University Press, 1993). The development of business interest in education embodies the pattern of increasingly imperial ideologies of management, as described by Charles S. Maier in "The Factory as Society: Ideologies of Industrial Management in the Twentieth Century," in *Ideas into Politics: Aspects of European History, 1880 to 1950,* ed. R. J. Bullen, H. Pogge von Strandmann, and A. B. Polonsky (London: Croom Helm, 1984), 147–63. Two Web sites provide a wealth of information about current business interest in early childhood education, including links to conference reports and research papers: see the Federal Reserve Bank of Minneapolis Web site, www .minneapolisfed.org/research/studies/earlychild/ and the site of the Committee for Economic Development, www.ced.org/projects/prek.shtml. The High/Scope study at the Perry Preschool in Ypsilanti, Michigan, is described, and the results reported, in *Significant Benefits: The High/Scope Perry Preschool Study through Age 27* (Ypsilanti, MI: High/Scope Press, 1993).

The role of computing in changing the organization and experience of work is central to the ongoing analysis of David F. Noble, particularly in "Social Change in Machine Design: The Case of Automatically Controlled Machine Tools, and a Challenge for Labor," *Politics and Society* 8 (1978): 313–47, and *Forces of Production: A Social History of Industrial Automation* (1984; rpt., New York: Oxford University Press, 1986). Among the works studied here Manuel Castells pays closest attention to the experience of work in a globalizing environment and its relationship to personal identity, in *The Rise of the Network Society*, 2nd ed. (Malden, MA: Blackwell, 2000). James W. Cortada's extensive survey of the changes in business wrought by computing is indispensable; see *The Digital Hand: How Computers Changed the Work of American Manufacturing, Transportation, and Retail Industries* (Oxford: Oxford University Press, 2004). Cortada has produced very useful bibliographic aids; see *A Bibliographic Guide to the History of Computer Applications, 1950–1990* (Westport, CT: Greenwood Press, 1996) and *Second Bibliographic Guide to the History of Computing, Computers, and the Information Processing Industry* (Westport, CT: Greenwood Press, 1996). His edited collection, *The Rise of the Knowledge Worker* (Boston: Butterworth-Heinemann, 1998) includes a number of contributions addressing the idea of a "new," or "post," industrial society. The how-to literature often reveals core motivations behind desired changes, as does Shoshana Zuboff's analysis of her findings as a consultant, described in an attempt to help managers guide the transition to an "infomated" workplace; see Zuboff, *In the Age of the Smart Machine: The Future of Work and Power* (New York: Basic Books, 1988). Important to considerations of efficiency in early computing was the cost and speed of tabulating or processing data. Arthur Norberg analyzes tabulators as important precursors to the electronic digital computer, in a truly mechanical sense, in "High-Technology Calculation in the Early 20th Century: Punched Card Machinery in Business and Government," *Technology and Culture* 31 (1990): 753–79. The influence of tabulation practices in the insurance industry is the subject of JoAnne Yates's studies "Co-Evolution of Information: Processing Technology and Use: Interaction between the Life Insurance and Tabulating Industries," *Business History Review* 67 (1993): 1–51, and "The Structuring of Early Computer Use in Life Insurance," *Journal of Design History* 12 (1999): 2–24. A very interesting consideration of labor and efficiency-boosting schemes through computing is Joan M. Greenbaum's *In the Name of Efficiency: Management Theory and Shopfloor Practice in Data-Processing Work* (Philadelphia: Temple University Press, 1979).

Page numbers in *italics* refer to figures.